# America's Environmental Report Card

© Reprinted with permission of King Features Syndicate.

# America's Environmental Report Card
## Are We Making the Grade?

Harvey Blatt

The MIT Press
Cambridge, Massachusetts
London, England

MIT Press books may be purchased at special quantity discounts for business or sales promotional use. For information, please e-mail special_sales@mitpress .mit.edu or write to Special Sales Department, The MIT Press, 5 Cambridge Center, Cambridge, MA 02142.

This book was set in Sabon by Graphic Composition, Inc.
Printed and bound in the United States of America.

Library of Congress Cataloging-in-Publication Data

Blatt, Harvey.
America's environmental report card : are we making the grade? / Harvey Blatt.
   p.   cm.
Includes bibliographical references and index.
ISBN 0-262-02572-8 (alk. paper)
1. United States—Environmental conditions. 2. Pollution—United States.
I. Title.

GE150.B58   2004
363.7'00973—dc22

2004040261

10 9 8 7 6 5 4 3 2 1

# Contents

# Preface

We didn't inherit this land from our ancestors, we borrow it from our children.
—Lakota Sioux proverb

I am located in Israel, and you may wonder why a foreign-based scientist would be writing about America's environment. Good question with a simple answer. I am a relative newcomer to Israel, having moved here from my lifelong home in America in 1994 after retiring from decades of teaching geology in American universities. One of my concerns as an American had been the nation's health, both figuratively and literally: figuratively in terms of dwindling oil and gas supplies and its effect on America's economic health, and literally in terms of the effects of pollution by oil and gas on human health.

Since coming to Israel, my environmental interests have broadened because of this country's chronic water shortages, pollution problems, lack of landfill space, agricultural difficulties, predicted increased aridity because of global warming, and relative nearness to Chernobyl. Israel was upwind from Chernobyl and suffered no ill effects from that disaster, but it certainly spiked my interest in the effects of radiation on living organisms. However, the focus in this book is on America's problems. I am still more familiar with these and feel I might be able to help its citizens understand the causes and possible solutions to the nation's environmental ills.

As the seventeenth-century cleric/poet John Donne said, "No man is an island," and this certainly applies to the publishing of books. The list of people who are part of the process includes the author, numerous manuscript reviewers, several types of editors, designers, and production people at the publishing house, and last but certainly not least, you the reader.

Without readers, the publishing of books would wither. So this book is dedicated to you, the reader, in the hope you will find its contents interesting and useful. I hope you will feel better informed about America's environmental difficulties when you finish it.

Harvey Blatt
*Jerusalem, Israel*

# Introduction

Environmentalists make terrible neighbors but great ancestors.
—David Brower

America's environment is in danger. According to public opinion polls there is rising concern about many kinds of environmental deterioration. High on the public's list are water pollution, toxic waste, air pollution, global warming, and radiation from nuclear power plants (particularly after Chernobyl). In a Gallup Poll in 2001, 75 to 81 percent favored setting higher emission and pollution standards for business and industry, setting higher auto emission standards for cars, more strongly enforcing federal environmental regulations, and spending more federal money on developing solar and wind power. A majority were opposed to expanding the use of nuclear energy. Protecting the environment was given priority over economic growth, 57 percent to 33 percent. The public's wishes are clear. But they are not optimistic about their desires being fulfilled. In 2000, 72 percent of public school parents polled believed the environment would become dirtier during their child's lifetime. An astonishing 95 percent of all adults want environmental education taught in our K-12 schools. The intensity of the American public's concern about environmental deterioration is perhaps best shown by a 2002 Gallup poll that indicated 63 percent of us would even be willing to roll back President George W. Bush's 2002 tax cuts to protect the environment. Now that's real concern!

The scientific community is also concerned about our incessant downgrading of the environment. However, the public and the scientific community do not always agree on which deteriorations are most serious. Part of the disagreement stems from how we define *risk*. Scientists, engineers, and other experts in the evaluation of hazards tend to use and interpret the term *risk* in a narrow actuarial sense—that is, as average annual mortality rates for the population. How many people are killed each year in floods? By how many years is the average American's life shortened by air pollution? What are the chances of getting cancer from an average lifetime number of X-rays by your doctor (1 in 700, between 0.1 and 0.2 percent)?

The public, on the other hand, commonly interprets *risk* in a very personal way, depending on whether they believe that they or their families are exposed. If you live along the Mississippi River, floods are a continual environmental concern for you and your family (about 110 people are killed each year). But if you reside in Nevada, the storage by the federal government of highly radioactive nuclear waste at Yucca Mountain is a much more pressing concern. We are all concerned about data suggesting that toxic agents in the environment have reduced the average male sperm count by 42 percent in the past 50 years and that estrogen-mimicking chemicals in drinking water can have a feminizing effect on organisms. Clearly, the concerns of the scientist and the public overlap. For example, both groups are concerned about the dangers of medical X-rays.

Can you believe what I say in this book? I hope so. A poll to determine who the public believes as sources of information revealed an extraordinary lack of trust in the objectivity of evaluations by the government, industry, media, trade unions, and religious organizations, which are widely perceived as having "an ax to grind," but scientists and environmentalists fared much better. However, friends and family members were considered almost 50 percent more believable than scientists (figure I.1). Unfortunately, your friends and close relatives will probably not write a book like this.

The topics we will consider in this book take into account both the scientists' concerns based on their actuarial way of thinking and the more personal evaluations of the general public. We will deal with water pollution in chapters 1 through 4, which treat water supply and pollution, flood dangers, water pollution by buried garbage in our town's landfill, and pesticide runoff in irrigation water from farms. The ways we generate the energy that powers our industrial society are explained in chapter 5, and the resultant global warming is discussed in chapter 6, which deals with carbon dioxide and the other heat-trapping gases emitted from our factories and cars. Chapter 7 is concerned with another result of the sources of en-

## People We Trust

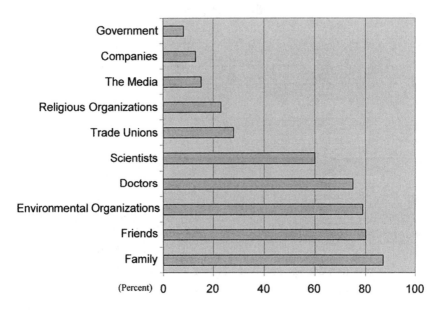

Figure I.1
Whom do you believe? (C. Marris and I. Langford, Who Do We Trust? *New Scientist,* September 28, 1996, p. 38). Reprinted with permission of *New Scientist.*

ergy we use, the filth in the air we breathe. The importance of ozone, its relationship to skin cancer, and the disastrous development of the "ozone hole" are dealt with in chapter 8. A lengthy chapter 9 is concerned with the problems of nuclear energy and the storage of its radioactive by-products. In chapter 10 I try to chart a realistic path to a sustainable future, one with enough water, clean and abundant soil, clean sources of energy, a stable climate, and pollution-free air.

I have tried to make the topics discussed in the book as accessible as possible. The tone is conversational, I have tried to sprinkle humor and entertaining anecdotes throughout the text, and I have included citations in each chapter so the reader can check my statements against statements by professionals in each field.

The environmental topics discussed in this book are those that I believe are viewed as most important to most Americans at the present time. There are others that I considered, such as the ever-increasing noise pollution that has damaged the hearing of many, perhaps most, Americans. The ongoing decimation of ecologically sound forests and their artificial replacement by fast-growing pine trees is another serious problem. The plundering of nearshore oceanic fisheries is an international disgrace that, if not stopped soon, may have disastrous effects on our food supply. Loss of biodiversity is a topic of worldwide concern. Desertification is a growing problem in some regions. Overpopulation is a growing concern. The list of insults to the environment is endless, but a line had to be drawn somewhere to prevent the book from becoming an encyclopedia.

Many of our environmental problems exist because for the first 80 percent of America's history the Europeans who settled here purposefully modified the environment for their needs with not enough concern for the long-term damage they were doing to the surroundings. The population was low and new land to develop seemed never-ending. Factories were located next to rivers, so their unwanted by-products could be dumped into them to be carried away and out of sight. Out of sight, out of mind. Unfortunately they have forced their way back into our minds as pure drinking water has become harder to find. Sales of bottled water have skyrocketed, partly in response to a public perception that public water supplies are not always safe.

Forests were believed to be so abundant that they were thought of as indestructible and were decimated to provide clear areas for farming and wood for houses and factories. Who thought that tree cutting increased flooding? The record-setting disastrous floods that hit the Midwestern part of the country in 1993 drew our attention to the need for costly disaster relief for millions of Americans. We all pay for this relief through our tax dollars.

What about the incredible volumes of garbage we produce? Take, for example, plastics. A glance almost anywhere in the modern world reveals that plastics are everywhere. Perusal of supermarket shelves, automobile bodies, styrofoam cups, or racks in dry-cleaning establishments reveals how indispensable these petroleum-based products have become. What happens to empty plastic soda bottles, car bumpers, or the filmy plastic that protects our newly cleaned suits and dresses? We dump it somewhere, usually in the trash bucket, from where it is most likely buried in a sanitary landfill, formerly called a garbage dump. In terms of human life span, plastics last forever and we are running out of conveniently located, usable dump space. What can we do?

America's agricultural abundance is the envy of the world, which depends heavily on our bounty. But this bounty is thought to have come largely as a result of the intensive use of pesticides that cause harm to us and other living things, as well as to the pests the chemicals are intended to control. Can our productivity be maintained without using these chemicals? Can contaminated soil be cleansed? Should organic farming be the wave of the future for American farmers? And what about genetically modified foods? Are they safe?

The lowland areas of the East and Gulf coasts are regularly swamped by hurricane waters, an event that may become more common as global warming causes glaciers to melt and sea level to rise. Much of the American population is clustered along the Atlantic and Pacific coasts, and by 2010 the coastal population is expected to reach more than 100 million, about one-third of all Americans. Can flooding within and at the fringes of the United States be controlled?

The world is getting warmer and more humid, in part because of the carbon dioxide we continually pump into our air. The United States produces 21 percent of the fossil-fuel-related carbon dioxide entering the air each year. How can we stop this change in our climate? The answer is clear. Stop burning coal and oil, the sources of nearly all the carbon dioxide increase. But can American industry survive without coal and oil? Can solar energy, wind power, and other renewable, inexhaustible, and nonpolluting sources of power replace coal and oil? If so, how soon, and why can't it be done "overnight"? After all, when we and the other major industrialized countries recognized that chemicals called chlorofluorocarbons (CFCs) were destroying our ozone shield about 20 years ago, we agreed to phase out their production and found substitutes. Can we do this with coal and oil?

Each day each of us inhales about 2,500 gallons of air. Along with the air we also suck in fumes from automobile exhaust, smog, lead, asbestos, and microscopic pieces of the rubber tires on our cars. The tread disappears into your lungs and accumulates there. No place on earth has pure, clean air nowadays. Some places are better off than others, however. Where are these places, and why is their air better than it is where most

Americans live? Can dirty air be cleansed and how? Should the government tax those who breathe clean air to finance cleaning up the lungs of citizens who suffer from air pollution?

These are some of the many environmental problems Americans are now being asked to think about and do something about. Some of our actions must be as individuals. We can use less water without feeling the pinch. So can farmers. The amount of plastic wrapping on the products in American stores is a national scandal and can painlessly be reduced. Wrapping can be a factor in our purchasing decisions without it causing damage to the products we want. People can choose not to build in flood-prone areas.

Some of our actions must be corporate. Surely laws can be passed to prevent the dumping of poisons into the water supply without destroying American industry. Genetic engineering and organic farming hold the promise of a largely pesticide-free agriculture. Communities can vote for small increases in their utility bills to help finance construction of pollution-free energy sources. Our future environment is in our hands. We have permitted this pollution to occur, and we can stop it if we choose to. In the chapters to follow we will consider our major environmental problems and discuss possible ways of solving them. And they must be solved eventually. Sooner or later the water must be purified, the air must be cleansed, and the garbage must be disposed of. The emphasis throughout will be on workable solutions that inflict minimal hardship on us all. After all, no one likes to sacrifice a high standard of living or abundant conveniences. Not even the author of this book!

# America's Environmental Report Card

# 1

## Water: Is There Enough and Is It Drinkable?

Men work on earth at many things;
Some till the soil, a few are kings;
But the noblest job beneath the sun
Is making Running Water run.
—John L. Ford, *Water and Wastewater Engineering*

Few of us think regularly about water. It seems limitless because it falls from the sky year after year. We turn on the tap and fresh, pure water comes out. Most of us have never known it to be otherwise. But problems that water specialists saw on the horizon many decades ago are now with us. Water shortages are a well-known problem not only in desert areas such as Tucson, Phoenix, Las Vegas, and Albuquerque but also in moister places like New York City. As America's population continues to grow at its current rate of 3 million each year, water shortages will creep into other large cities as well. Adding to the water problems caused by increasing numbers is the increasing concentration of our population in cities. The nation's population is increasing about 1 percent per year and the growth of cities is much faster. About three-quarters of all Americans now live in large cities. That is where most jobs and growth opportunities are.

Perhaps even more frightening than the looming shortage of water is the amount of impure water we are drinking. Despite marked improvement since passage of the Clean Water Act 35 years ago, the United Nations estimates that 5.6 million Americans (2 percent of us) drink water that does not meet safety standards. Chemical contaminants are present in all our major streams and in 90 percent of our underground aquifers. Twenty-four percent of Americans refuse to drink tap water. Sixty-five percent take such precautions as treating water in their homes by filtering or boiling it.

More than half of us drink bottled water despite a 1997 UN study that showed bottled water was in no way superior to New York City tap water. And the Natural Resources Defense Council in 1999 estimated that at least 25 percent of bottled water is in fact ordinary tap water. One bottled-water supplier was found to be drawing its water from a well in the middle of an industrial parking lot next to a hazardous waste site![1]

There probably is more than one reason more than half of all Americans drink bottled water. Not only suspicions about our city's water may be involved. Thanks to advertising, there is a certain cachet or possibly snob appeal to imbibing a glass of Perrier or Evian imported from France rather than the liquid the city supplies. But whatever the reason, bottled water is the fastest-growing major beverage category in America. On average, each of us in 2000 drank 53 gallons of bottled water (table 1.1). Sales have increased ninefold in the past 20 years, tripled in the last 10, and increased 30 percent between 2000 and 2001 and 11 percent more in 2002, despite the fact that bottled water costs 120 to 7,500 times more than tap water and 6 times more than gasoline.

Even our treasured pets can enjoy the thrill of bottled water designed especially for them. The K9 Water Company in California (of course, where else?) sells beef-, liver-, chicken-, and lamb-flavored bottled water for dogs. You can even get all four in a combo pack "so your dog can decide . . . "

**Table 1.1**
Beverages Americans drink in gallons per year (International Bottled Water Association)

| | |
|---|---|
| Water | 140 |
| Tap | 87 |
| Bottled | 53 |
| Carbonated soda | 59 |
| Coffee | 46 |
| Juice | 43 |
| Milk | 39 |
| Tea | 27 |
| Alcoholic drinks | 23 |
| New age beverages | 14 |
| Sports drinks | 11 |
| TOTAL | 402 |

Pollution from farms, factories, and even the pipes that bring the water to our homes is increasing. Underground water supplies in about half the states have been contaminated with hazardous wastewater legally injected into it. The wastewater came from chemical plants and other industrial sources that produce materials essential to the way we live.

Lead in drinking water is a problem in many major cities, including Chicago, San Francisco, Boston, New York, and Washington, our nation's capital. Water pipes in buildings built before 1960 were made of lead, and lead solder to seal water pipes was in use until 1988. It is uncertain how many Americans have health maladies caused by ingesting lead. Lead causes brain damage, among other maladies, but we do not know whether the lead in Washington's drinking water, where 16 percent of the water pipes are made of lead, has affected legislative judgment in recent Congresses. The way we waste and contaminate our water supplies has generated a new word— *hydrocide,* patterned after the more familiar word *suicide.*

## The Water Cycle

The journey of water is round, and its loss, too, moves in a circle, following us around the world as we lose something of such immense value that we do not yet even know its name.

—Linda Hogan, *Northern Lights*

Most of America's large cities use surface water. The amount of surface water available to each American for all purposes (personal, industrial, agricultural, and so on) from rainfall, rivers, and lakes is 138,000 gallons per day. This number is the result of a system of water circulation known as the *water cycle.* Pure water is evaporated from the salty ocean, is carried by winds over the land surface, and as air temperatures and land elevations change, the moisture is dropped from the air onto a thirsty population. Most of this moisture falls on land, runs off into streams and rivers where it is available for our use, and eventually finds its way back to the ocean. Some of this heaven-sent moisture is taken directly into plants and combined with carbon dioxide gas from the air to produce plant tissues (biomass). Some of the precipitation soaks into the soil and continues downward hundreds or thousands of feet into empty spaces in the underlying rocks. This becomes an underground water supply known as *groundwater*. Some precipitation falls directly into lakes, such as the Great Lakes that form part of the boundary between the United States and Canada. And some of the moisture that falls to the ground evaporates back into the moving air before it can flow into streams, enter lakes, or soak into the ground. When all these gains and losses are totaled up each of us ends up with a theoretical 138,000 gallons per day to spend as we see fit, for drinking, growing crops, manufacturing steel, or flushing toilets.

The expression "each of us" is, of course, a statistical average. Obviously, some of us end up with more than others. If you live in the Western half of the country you average less than your "fair share," perhaps 20 inches of rain and snow a year. If you live in the Eastern half, your cup runneth over with perhaps 40 inches a year (figure 1.1). Life is not fair. Neither is the distribution of water. But we must deal with the world as nature provides it. How do Americans deal with it? The answer is "very wastefully." All of us contribute to the national hydrocide. Consider the following facts.

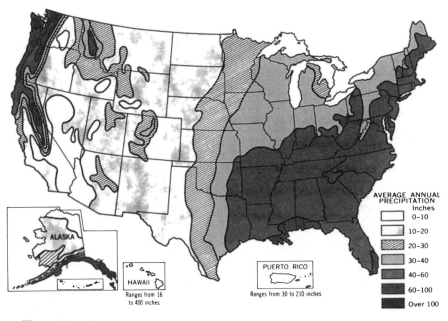

**Figure 1.1**
Average annual precipitation in the United States (U.S. Water Resources Council, 1968, *The Nation's Water Resources*).

• The channel of the formerly mighty Colorado River in the Western United States is dry when it reaches its outlet at the Gulf of California, the result of too much removal by users upstream.

• Water is being removed from underground reservoirs many times faster than it is naturally replenished. Most of the withdrawals from our underground water bank are for crop irrigation.

• Water usage in the United States has increased sixfold since 1900 although the population has increased less than fourfold.

• One in five Americans drinks water from a treatment plant that violates safety standards. Forty percent of these plants release water with dangerously high levels of disease-causing bacteria.

• According to the U.S. Geological Survey, the water in 47 percent of city wells contains toxic organic compounds. In rural areas, 14 percent of wells contain these chemicals. Over a person's lifetime, ingestion of these chemicals has adverse health effects such as cancer and reproductive problems.

Let's look at these factors in our national hydrocide to see why we have them and what we might do to remedy them.

### The Colorado River

The first thing they noticed was that the river was no longer there. Somebody had removed the Colorado River.
—Edward Abbey, *The Monkey Wrench Gang*

The dry channel at the southern end of the Colorado River is perhaps the prime example of surface-water scarcity produced by human activities (figure 1.2). The river originates in western Colorado, then flows through southeastern Utah and along the boundary between California and Arizona before entering Mexico and spilling into the Gulf of California. Actually, the word *spilling* is inaccurate, because the river channel is dry at its contact with the Gulf. Humans are to blame. The Colorado River is among the most heavily plumbed rivers in the world, providing water for 30 million people, one-tenth of the American population.

The region through which the river flows is semiarid, with an annual precipitation of only 15 inches, and nearly 90 percent of this moisture evaporates before reaching the river channel. Even so, the average volume

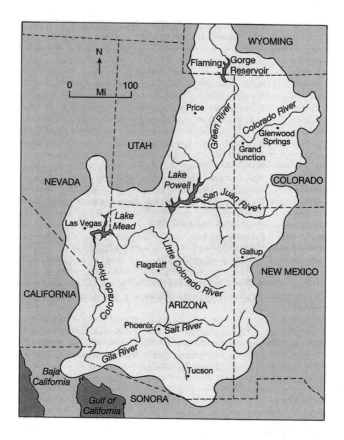

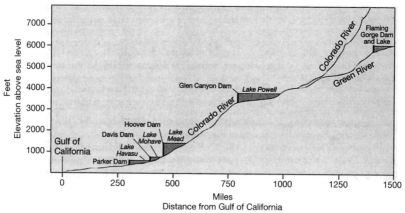

**Figure 1.2**
Drainage area served by the Colorado River and the dams constructed to minimize
annual variations in rainfall.

of water carried by the river is more than 15 million acre-feet per year. (An acre-foot is an area of 1 acre covered with water to a depth of 1 foot; about 325,000 gallons of water, enough to supply the water needs of a family of 5 for 1 year.) You would think that 5 trillion gallons of water a year (15 million × 325,000) would be enough to keep people happy. And it was, until the increase in the number of people in southern California in the first half of the twentieth century and the growth of Las Vegas, Phoenix, and Tucson in the last half.

The influx of people to California and Arizona quickly generated water problems. Southern Californians wanted to transport Colorado River water westward to supplement the inadequate amount of precipitation that nature supplies to Los Angeles, about 15 inches per year. Phoenix, with only 8 inches of rainfall a year but with a climate that appeals to retirees from the frigid Northeast, also wanted Colorado River water. And there was a growing agricultural base in central Arizona that depended on water for irrigation. Also wanting more water after World War II was arid Las Vegas, with only 4 inches of rain per year. This city's heady mixture of gambling and prostitution stimulated its growth from a small community in 1950 to its position today as one of America's fastest-growing metropolitan areas, with a population of 1,500,000.

What should be done? Who owns the water, anyway? Legal battles over the ownership of Colorado River water brew continuously among the states that border the river and also between the United States and Mexico, because the water has been overappropriated. More water has been allocated to the states than the river can supply. Problems began in 1922 when an agreement called the Colorado River Compact divided the river into an upper and a lower part. Wyoming, Colorado, Utah, and New Mexico were to share the water of the upper part of the drainage area, and Nevada, Arizona, and California were to share the lower-basin water. The users of each part were allocated 7.5 million acre-feet per year, half the average yearly flow of 15 million acre-feet. In 1922 the yearly flow was higher than normal at about 20 million acre-feet, a heady surplus. This agreement among the states was followed in 1945 by a treaty with Mexico that allocated our southern neighbor a minimum of 1.5 million acre-feet per year. Hence, more water was allocated in these two agreements than the river contains in an average year.

As we know, an average is a central value around which there is variation. In rainy years river flow will exceed the 15 million acre-feet average; in dry years it will be less. Recall that the treaties allocated acre-feet of water to the contestants, not percentages; the allocation was 7.5 million acre-feet, not 50 percent of the annual flow. Since 1922 or 1945 there have been many years of less-than-average flow. In 1934 flow was less than 5 million acre-feet; in 1940 it was 7 million; and in 1963 and 1964 flow was a minuscule 3 million acre-feet. We have entered lawyers' heaven.

The method chosen to circumvent these unfortunate allocations was to build dams along the river, which would store water during wet years and release it during dry years. This would smooth out the yearly variations in river flow. There are now ten major dams along the Colorado River. Obviously, the dams could not change the predam average yearly flow of 15 million acre-feet. They could only make the yearly variations in rainfall less traumatic.

The phenomenal population growth in southern California, Arizona, and southern Nevada has drawn increasing attention to the inadequacy of the water supply in this region. Colorado River water provides for the households of tens of millions of people, fills swimming pools and sprinkles green lawns in Los Angeles, powers neon lights in Las Vegas casinos, and irrigates 2 million acres of farmland in southern California, southern Arizona, and northern Mexico. Turbines in the dams also generate nearly 12 billion kilowatt-hours of electricity annually. There is no long-term solution to the problem of inadequate surface water for the burgeoning population in this area of the United States.

## Subsurface Water

Humans build their societies around consumption of fossil water long buried in the earth, and these societies, being based on temporary resources, face the problem of being temporary themselves.
—Charles Bowden

What about water located underground? More than half the U.S. population depends on subsurface water as their primary source of drinking water. It has been estimated that the amount of freshwater contained in rocks below the ground, estimated to be 33 quadrillion (33,000,000,000,000,000) gallons,[2] is about 100 times the amount held in freshwater lakes and rivers. Perhaps this is where we should look for additional water. Where do large supplies of subsurface water (called groundwater) occur and how do we tap into it? Nearly all groundwater suitable for drinking or irrigation occurs within 1,000 feet of the ground surface in tiny holes in rocks. Most of these holes are in rocks called sandstones and limestones. The holes are called pores and the percentage of pores in a rock is called its porosity. Typical porosities in water-bearing rocks are 10–20 percent. Although pores in rocks are unimaginably abundant, most pores are very small, with diameters between 1/500 and 1/25 of an inch. Few of them can be seen without using a microscope, so most people are unaware of their existence. Underground water is not generally located in large caverns similar to Carlsbad Cavern or Mammoth Cave but in microscopic cavities in rocks.

A layer of rock that yields water in amounts large enough to be useful is called an aquifer. Aquifers must not only contain lots of water-filled pores, but the pores must also be interconnected (the amount of interconnection is called the permeability) so the water in the rock can move toward the wells that have been drilled into it. How much water can we expect to get from a suitable rock? Nearly all aquifers are layered rocks that are tens to hundreds of feet thick. They are miles to tens or even hundreds of miles in length and width.

However, it is never possible to withdraw all the water. Perhaps 20 percent will remain unrecoverable in the aquifer. There are hundreds of aquifers of various sizes in the United States and they supply 25 percent of America's total water needs. Groundwater wells supply about 37 percent of all "city water," about 96 percent of rural domestic supplies, and 34 percent of the water used in agriculture. We withdraw 28 trillion gallons from aquifers each year. However, like surface-water supplies, groundwater reservoirs can be overtapped. One well-studied example of an overdrawn aquifer is the Ogallala Formation.

### The Ogallala Aquifer
The body of sandstone rock called the Ogallala Formation is the largest and best-studied aquifer in the United States (figure 1.3). It has been a major supplier of water to the American midcontinent, from Nebraska southward

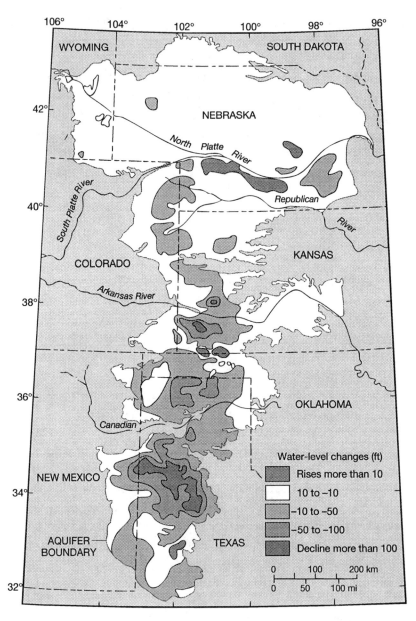

**Figure 1.3**
Changes in water level in the Ogallala aquifer between 1850 and 1980. The declines have continued to the present day (U.S. Geological Survey).

to Texas. Today, Ogallala water irrigates more than 14 million acres of farmland. It supplies water to 20 percent of all irrigated land in the United States. The aquifer extends over 225,000 square miles and holds more than 70 quadrillion gallons of water (70,000,000,000,000,000 gallons). It averages 200 feet thick but in some areas the thickness reaches 1,400 feet. This aquifer is truly a monster in size. Water in the Ogallala accumulated undisturbed from rainfall over millions of years, but for the past 80 years this water has been withdrawn at an ever-increasing rate. Without Ogallala water there would be little agriculture in this region because annual rainfall is only 16 to 20 inches, not enough to stimulate the agricultural abundance we have come to expect. Water from the Ogallala aquifer serves an area that produces about 25 percent of U.S. food-grain exports and 40 percent of wheat, flour, and cotton exports.

The aquifer can yield as much as 1,000 gallons per minute, 24 hours a day. But thousands of wells tap the Ogallala, so that the rate of withdrawal is currently eight times greater than the rate of replenishment by the low annual rainfall.[3] Without Ogallala water, agricultural production will drop to a third of its present volume. To date, only about 5 percent of the total groundwater resource has been used up, but water levels have declined 30 to 60 feet in large areas of Texas. Wells must be deepened and the energy cost to pump the water to the surface increases to the point where farming becomes uneconomical. In northeast Texas the area under irrigation dropped by one-third between 1974 and 1989 because irrigation from the Ogallala no longer is practical. If present usage continues, the Ogallala will be effectively dry within a few decades, with disastrous effects on the economy of a large area of the United States. Our present ability to irrigate at low cost is coming to an end, not only in the midcontinent but in other areas as well.

**Water Use**

To take anything for granted, is in a real sense, to neglect it and that is how most of us treat water.
—Robert Raikes, *Water, Weather, and Prehistory*

What part of the American economy is responsible for our dwindling water supply? Where can the biggest cuts be made? Is anyone trying to make these cuts? How can we help?

## Agriculture

Probably the chief reason water usage grew three times faster than the growth in population since 1900 is the expansion of agriculture. Agriculture is by far the biggest consumptive water user in the United States, most of it groundwater (figure 1.4). Agriculture accounts for 43 percent of our water use. Surprisingly, the most productive croplands are located in areas with relatively low annual rainfall. The average yearly precipitation for the United States is 30 inches per year. The San Joaquin Valley of California yields 50 percent of the nation's fruit and vegetables but has only 8–12 inches of annual rainfall. The Midwest produces most of our grain but has 10–30 inches of precipitation, marginal for farming.

RELATIVE PRODUCTION OF GROUND WATER, 1980
— In Millions of Gallons Per Day.

**Figure 1.4**
Where groundwater use is concentrated. The greatest use by far is in the major agricultural areas (U.S. Water Resources Council, 1980).

Our agricultural abundance in these water-deficient areas has been achieved in two ways. The first is the enormous government water subsidy to farmers. For example, in California's San Joaquin Valley farmers can buy a thousand cubic meters (264,200 gallons) of water from a federal project for $2.84, even though it cost the government $24.84 to deliver that water, nine times as much. In terms of the farmers' profit, the water is actually worth $80–$160. The second way farmers survive in the Valley is by supplementing inadequate rainfall with irrigation water, most of it from groundwater. Can we cut the amount of water used for irrigation without affecting the amount of food we produce? The answer is yes; water use can be reduced significantly. In many agricultural areas it has been and the result is that although population increased by 40 million since 1980, the nation used 10 percent less water in 1995 than in 1980.

How has the reduction in agricultural water use been accomplished? By improved irrigation methods, an improvement that greatly increases the amount of water available for other needs. A modest 15 percent efficiency gain in irrigation frees up double the amount of water used by humans for other purposes. About half of America's cropland is irrigated using large sprinklers that spray into the air about 10 feet above the ground. With the help of the wind, the water is distributed over a wide area. This method of irrigation is relatively inefficient because much of the water evaporates without hitting the ground. A newer sprinkler design delivers water closer to the crops by means of drip tubes extending vertically from the sprinkler arm. Efficiencies as high as 95 percent have been reported. Adapting an existing sprinkler for this system costs about $25–$65 per acre, and the water, energy, and crop-yield gains typically make it a cost-effective investment, recouped in 2 to 4 years. Improvements in efficiency such as the drip sprinklers have reduced depletion of the Ogallala aquifer in the Texas High Plains by 30 percent.[4]

Another very efficient method of getting needed water to crops and getting more crop per drop is drip irrigation, used on only 4 percent of our irrigated cropland. Almost all of the water reaches the plant; efficiencies with this method are more than 90 percent. Losses of water to evaporation and runoff are nearly eliminated. Water use is reduced by 30–70 percent and crop yields are increased by 20–90 percent over standard irrigation methods.

But a drip irrigation system is expensive to install. Miles of pipes and tubes must be laid on the rows of plants, and the holes in the pipe through which the water drips onto the roots of the plant should be as close to the plant as possible. Installation of a drip-irrigation system costs about $1,000 per acre. Perhaps the federal government could offer tax incentives to encourage large farms to switch to drip irrigation. Even without a tax incentive, increases in the irrigation efficiency of American agriculture will have to be made. The choice is "change or die." Groundwater reserves are being depleted almost everywhere.

**Industry**

Industry accounts for 38 percent of America's water use. The bulk of water used in manufacturing occurs in four industries: paper, chemicals, petroleum, and metals. One or more of these industries is involved in the production of most of the products we use every day, from clothes and computers to cars and plastics. All require large amounts of water to produce. According to the U.S. Geological Survey, producing 1 pound of paper uses about 100 gallons of water. Making a ton of steel requires 50,000 gallons; aluminum, 1,000,000 gallons.

However, in contrast to the water used in agriculture and in the home (see below), only 10 percent of industrial water is actually consumed. Nearly all of it is used for cooling, processing, and other activities that may heat or pollute the water but do not use it up. This creates the possibility of recycling water within a factory and many industrial operations take advantage of this opportunity. More than 95 percent of the water used for steel production and processing is recycled. Intensive recycling of water by American industry has reduced its water use by 36 percent since 1950, while industrial output has nearly quadrupled. Whereas our manufacturing operations were using each gallon of water supplied to them an average of 1.8 times in 1954, the recycling rate is now about 17.

In deciding how much to recycle, a manufacturing plant balances the cost of getting water and treating it before disposal against the cost of adding equipment to treat and reuse wastewater within the plant. In most industries, recycling partially offsets its costs by recovering valuable materials, such as nickel and chrome from plating operations, or fiber from the manufacture of paper. Studies have shown that industrial use of water per unit of production has steadily declined in recent decades. No doubt much of the decline has resulted from passage of the Clean Water Act in 1972, which restricts the discharge of untreated wastewater. As the cost of obtaining water and treating it after use continue to rise, recycling becomes increasingly more cost-effective.

## Home Use

About 19 percent of the nation's water use is in the home, so that part of the reason for our diminishing water supply lies in increased cleanliness and the nearly universal access Americans have to modern plumbing. In 1900 less than one in five homes had running water; today nearly all homes do. Three-quarters of the water you use at home you use in the bathroom, mostly for showers and toilet flushing (figure 1.5).

**Showers** In 1900 Americans bathed or showered only once or twice a week (or less!). Most women washed their hair only once a month (and used borax or egg yolks for shampoo). Only 14 percent of our homes had bathtubs. As late as 1950 only 29 percent of Americans bathed daily in the winter; in 1999 it was 75 percent. In many parts of rural America, bathing in the early 1900s was often more a seasonal than a daily affair. The notion of being wet all over at once, indoors, was little short of revolutionary and

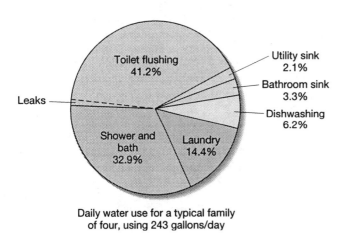

Daily water use for a typical family
of four, using 243 gallons/day

**Figure 1.5**
Use of water in an average American home (American Water Works Association).

workingmen might prefer to walk to a river for the privilege of cleanliness. Nowadays, most of us shower daily, a water drain of at least 5 gallons for every minute the water is running. The feel of a massaging hot shower for 10 minutes may be invigorating, but when 295 million people do it every day the water use is staggering.

No one expects Americans to stop showering, or even to shower less frequently. But our use of shower water can easily be reduced significantly. One way is to install a plastic or metal washer behind the showerhead to restrict the flow of water. Low-flow showerheads can also be purchased for a few dollars. They cut the showerhead volume by 50 percent, saving about 5,000 gallons of water per person over a year. Multiply by 295 million to see the nationwide annual saving. A cost-free way to cut use of shower water is to step into the shower, wet yourself, turn off the water, soap yourself, and then rinse the water-sweat-dirt mixture from your beautiful frame. Three minutes of running shower water are enough to accomplish the body-cleaning job, be your body large or small. Keep in mind that in addition to the greatly increased use of showers by Americans, there are a lot more of us taking showers. In 1900 there were only 76 million of us; today we are 295 million. And the amount of water that falls on the 50 states has increased only slightly (5–10 percent) during the last hundred years (a result of global warming).

**Toilets**    Although the first flush toilet was developed more than 3,000 years ago, the concept seems to have been lost over the millennia, and human-waste disposal in 1900 was as primitive as it had been 2,000 years earlier. It consisted either of an outdoor privy (known in colonial times as a "necessary house") or a chamber pot, to be emptied into privy pits. Water closets, as they were named, began slowly appearing in America in the 1800s, imported from England. But adoption in this country was slow. In 1900 only 10 percent of American homes had a flush toilet. There were cultural concerns about performing indoors a process hitherto associated with nature. In addition, the cost of installation and the problem of disposal of the human waste kept them from the masses. The disposal problem was solved in the 1860s by Thomas Crapper (yes, that really is his name; his biography is titled *Flushed with Pride*), who commercialized the flush toilet. Today, 98 percent of American homes have at least one flush toilet, a facility each of us visits 2,500 times a year, about 6–8 times a day. And toilet enthusiasts have their own professional organization. The World Toilet Organization holds annual congresses highlighting toilet-related issues. Unbeknownst to most of us, there are toilet associations worldwide promoting toilet education and culture.

The toilet in use in most American homes until relatively recently used 5 gallons of water per flush. Sensing that too much water was going down the drain together with the other stuff, Congress in 1992 mandated that toilets sold in the United States use no more than 1.9 gallons per flush. The latest ultra low-flow models use only 1.6 gallons per flush, a water saving of 70 percent over the 5-gallon models. In 1988, Massachusetts became the first state to require that all new toilets installed use no more than 1.6 gallons. If we assume that an average person spending quality time in the bathroom makes five flushes per day, she or he will save 2,920 gallons of water in a year by using only 1.6 gallons per flush. Multiply by 295 million to see the nationwide saving. Many municipalities have started requiring low-flow toilets in new construction. And no wonder. American toilets flushed about 16.4 million times and used 48.5 million gallons of water *at halftime during the 1999 televised Thanksgiving Day football game.*[5] The mind swirls at the thought of flushes by the Super Bowl TV audience, estimated to have been 144 million in the United States for the 2004 event.

Ever count the number of flushes your family makes each day? Probably not. As an interesting experiment, put a notepad and pencil in each bathroom in your house and ask each person to keep track. Almost certainly the total will be higher than you think.

Even worse than the careless hand on the flush mechanism is the silent toilet-bowl leak, probably the single greatest water waster in most homes. It has been estimated that about 20 percent of all toilets leak. In 2002, water leaks accounted for 14 percent of home water use in the typical single-family home. In some areas of the country, such leaks cause about 95 percent of the complaints to city governments about excessive water bills. A leak of 1 gallon every 6 minutes—not an unusual amount—totals 10 gallons per hour or 240 gallons per day, almost equal to the average amount of water consumed each day in a single-family home. The leak nearly doubles total water consumption. To detect the silent leak in the toilet bowl, place a few drops of food coloring in the tank and wait 5–10 minutes; if the color shows up in the bowl, there's a leak.

**Other Bathroom Uses**    Another way we waste water in the bathroom is during hand washing, brushing teeth, and shaving. One hand wash, one teeth cleaning, and one shave with the faucet running uses 20–25 gallons of water. All these standard procedures can be accomplished with a small amount of water in a stoppered basin, cutting water use by 90 percent.

**Clothes Washing**    Few people had washing machines in 1900. Today Americans own 80 million of them. Eighty-one percent of American families have one. The average washing machine uses about 30 gallons of water to wash a full load and about 35 billion loads of wash are done in the U.S. each year. Fourteen percent of household water consumption is used in washing machines. Keeping us in clean clothes uses perhaps 1,000 billion gallons of water per year. However, newer washing machines use less water than older ones and new federal standards in 2002 for these gadgets ensure that this trend will continue. Obviously, Americans are not going to trash all their washing machines and use a less water-intensive method to clean dirty clothes, and no one is going to volunteer to wear smelly clothes to save water. So more efficient machines are the only realistic solution. However, we should recognize that the invention of the washing machine has greatly increased America's use of water since 1900.

**Dish Washing**    An automatic dishwasher is present in 57 percent of American homes, using 5–11 gallons of water per run. When dish washing is done by hand, a savings of at least 50 percent can be made by filling the kitchen sink for the soapy part of the process and conserving additional water by not leaving the faucet running during rinsing. A running faucet during washing and rinsing can use thirty times more water than an electric dishwasher.

**Car Washing**    Cars and trucks were rare sights in 1900. Today the average American adult owns *at least* one car and washes it with the hose running full blast for the 15–20 minutes it takes to sanitize our proudest possession. A ½-inch garden hose under normal water pressure pours out more than 10 gallons per minute; a ¾-inch hose delivers almost 32 gallons per minute. A better way to wash your car is to use a bucket with soapy water for the washing. Don't forget to turn off the hose when you finish rinsing. Should you or one of your children forgetfully leave the hose on overnight, you can easily waste twice as much water as your family uses in an entire month.

## Leaky Water Pipes

Finally, there is a big water waster that we, as individuals, can do nothing about. Many of the 880,000 miles of water mains in American cities are old (a century is not unusual) and leaky.[6] Water mains break 237,600 times each year in the United States, 0.27 breaks per mile of pipe per year. (Water mains are the central conduits through which city water is piped; pipes from the street curb to the homeowner's water tap are called service lines.) In New York City about 600 aging water mains break each year and the city loses 15 percent of its municipally pumped water to leaks, which is about the national average. Buffalo loses 40 percent. St. Louis's water system predates the Civil War. According to the American Water Works Association, a "huge wave" of water-supply pipes laid 50–100 years ago are approaching the end of their useful lives, and "we can expect to see significant increases in break rates and therefore repair costs over the coming decades." The EPA estimates that replacing these old parts of our metropolitan infrastructure will cost at least $138 billion. Local governments and ratepayers currently provide 90 percent of costs to build, operate, and maintain public water and sewer systems. A major federal investment is needed to close a $23-billion-a-year gap between infrastructure needs and present funding in order to meet priorities in the federal Clean Water Act and the Safe Drinking Water Act. As noted by the Water Infrastructure Network, "If we do nothing, the nation can expect increased threats to public health, environmental degradation, and real economic losses. At times and in places, these threats will be small and barely noticeable, but over the next two decades, and even more quickly in some locations, losses will mount and solutions will be financially unmanageable."[7] Obviously, these repairs and replacements will have to be made eventually, if not by you then by your children or grandchildren. Given the current extent of America's water usage and pollution problems, the sooner we begin replacement of the water distribution system, the better.

## Water Prices

There are more than 200,000 public water systems in the United States, and Americans greatly underpay in all of them for the water they use. The unrealistically low cost of public water supplies is a serious impediment to water conservation. An average urban family uses about 12,000 gallons per month, costing only about $25. At this ridiculously low price we can refill an 8-ounce glass of water with tap water 2,500 times for less than the cost of a can of soda. At such a low price for municipal water there is no financial incentive to conserve. If state utility commissions allowed utilities to double or triple their charges to reflect national water scarcity, conservation might become more popular. Researchers have found that domestic water use drops by from 3 to 7 percent when prices increase 10 percent.[8] Given human nature, conservation will not become more widespread until water shortages become more widely understood and felt in the wallet. An increase in price doesn't make me any happier than it does you. But there is no realistic alternative. Cheap water is not a birthright.

Fifty years of studies have shown that water demand is responsive to price changes, both in the short term, as individuals and companies respond by making do with less, and in the long term, as they turn to more efficient devices in the home and workplace. For example, when Boulder, Colorado, moved from unmetered to metered systems, water use per person dropped by 40 percent and stayed there.

Water is not only underpriced; it's also inappropriately priced.[9] Most of the 60,000 water systems in the United States charge uniform rates, meaning that consumers pay the same rate per gallon no matter how much they use each month. One-third of municipalities use an even worse pricing method. They offer volume discounts; the more water you use the less you pay. Only 22 percent of utilities charge higher rates for those who use more. And less than 2 percent of water companies charge more during summer, when demand is greater. To avoid hurting the poor, water utilities can follow the example of electric utilities that subsidize the first kilowatt-hours of electricity use with very low "lifeline rates." At some point we will have to change the extravagant way we use water. The patient is sick and getting worse. It is time to do something.

An innovative approach to conserving the nation's water and homeowners' money has recently been introduced in Brazil: digital water.[10] Brazilian engineers have come up with an electronic device known as a water manager. With this device customers draw water on a strictly pay-as-you-go basis. The water user buys a smart card at a local convenience store that, like a long-distance telephone card, is programmed for a certain number of credits. At home, the purchaser punches the card's code into a small keyboard and pushes the LOAD key. The water manager automatically sends a signal to the water company to supply you with water. When the user runs out of credits, just push the LOAN key and the water authority will pump you a bridge loan to carry you until you can run out and purchase another card. According to Brazilian officials, water managers save water, electrical power, and money. They discovered that households using the water manager saved 40 percent on their water bills. Becoming increasingly conscious of what something costs gets people to use less.

## Recycling Wastewater

The bad news is that if the drought keeps up, within a few years we'll all be drinking reclaimed sewer water. The good news is that there won't be enough to go around.
—Bill Miller

Although recycled wastewater still totals less than 1 percent of America's water use, the amount is increasing rapidly. Hundreds of American cities are using recycled water for nondrinking purposes such as crop irrigation and landscaping. California and Florida, our major fruit and vegetable producers, have wholeheartedly embraced the practice of irrigating crops and public areas with treated municipal wastewater.

Other nonpotable applications include cooling water for power plants and oil refineries, industrial-process water for such facilities as paper mills, toilet flushing, dust control, construction activities, concrete mixing, wetland enhancement, and artificial lakes.

Many communities are studying the safety, economics, and feasibi of directing treated sewer water into the ground to replenish dwindl aquifers, even those tapped for public drinking water. This practice is n federally regulated, but all water used for drinking or crop irrigation mu meet EPA purity standards. In other words, you can inject what you war but when you draw it back up to use it again it must meet safety standards The nation of Israel is a leader in the use of purified recycled wastewater. The government projects that one-third of its water needs in 2010 will be met by reclaimed and recycled sewage water.

## Water Pollution

Water, water, everywhere, nor any drop to drink.
—Samuel Coleridge

Americans are making a two-pronged attack on their water supply. Not only do we use it extravagantly but we pour harmful chemicals in it as well. In his research for a PBS documentary report, Bill Moyers found out that his blood contains 84 synthetic chemicals. He is in relatively good shape. The bodies of most people on earth contain traces of some 500 synthetic chemicals that didn't exist before the 1920s.[11] How many of these are harmful and in what amounts is largely unknown. Basic toxicity data are not publicly available for nearly 75 percent of the 3,000 chemicals produced in the highest volume each year, excluding pesticides.[12] Are we crazy? Who is pouring poisons in our water and why?

### Scary Indicators

A nationwide reconnaissance of pharmaceuticals, hormones, and other organic wastewater contaminants in 139 U.S. streams in 30 states was conducted by the U.S. Geological Survey in 2001.[13] They searched for 95 chemicals and found one or more of them in all 139 streams sampled. A mixture of 7 or more were found in half the streams. Most of the contaminants were present in concentrations that did not exceed current drinking-water guidelines, but recent studies indicate that mixtures of certain chemicals may produce greater than anticipated effects—that is, more severe symptoms, unpredicted effects on organs not known to be affected by the individual components, and effects at concentrations much lower than those known to be harmful for the individual components.[14] Therefore, concentrations of individual chemicals in a mixture to which a person is exposed are not necessarily indicative of the ultimate effects.

Late in 2002 the H. John Heinz III Center for Science, Economics, and the Environment released the results of a 5-year study of the nation's streams and groundwater.[15] It revealed that 13 percent of the streams were seriously polluted, as were 26 percent of the groundwater samples.

As of 2003, 270,000 miles of rivers and streams are too polluted for fishing and swimming.[16] In 1975 a health advisory was issued (still in place in 2003) that children and women of childbearing age should not eat fish from the 315-mile-long Hudson River in New York because of pollution by a cancer-causing chemical. In 1984, 193 miles of the river was declared a Superfund site. Cleanup is expected to take about 6 years. The Environmental Protection Agency reported in 1998 that 40 percent of America's rivers, lakes, and estuaries were no longer suitable for fishing and swimming, largely due to runoff of polluted water from agricultural and urban areas.[17] Forty-one states now advise anglers to limit wild-fish consumption because of contamination by mercury. Bass are particularly adept at accumulating mercury in their tissues. All eight states bordering the Great Lakes restrict consumption of fish from the lakes because of the high concentrations of mercury, pesticides, and more exotic chemicals such as dioxins and PCBs in the fish tissues. Children, whose bodies are growing rapidly, are particularly sensitive to these pollutants. The list of diseases caused by high levels of these pollutants reads like a medical dictionary.

In the Everglades, a sign posted by the National Park Service reads:

WARNING. HEALTH HAZARD
Do not eat more than one bass per week per adult due to high mercury content. Children and pregnant women should not eat bass.

In July 2001 Massachusetts public health officials warned young women and children under 12 to stop eating most fish caught in state rivers and lakes because of mercury poisoning, and to avoid some other seafood. According to the Centers for Disease Control and Prevention in 2001, one of ten American women of childbearing age is at risk for having a baby born with neurological problems due to in utero mercury exposure. That's 375,000 babies at risk every year. Most of the mercury comes from America's coal-fired power plants.

The Mississippi River is the most polluted river in the United States.[18] The river's banks are lined with city-sized chemical plants, which dump more than 50 million pounds of toxins annually. The 150-mile stretch from Baton Rouge to New Orleans is known to water-pollution specialists as the "Cancer Corridor." This section of America's largest waterway contains 500 hazardous-waste sites and the highest concentration of manufacturers, users, and disposers of toxic chemicals in the United States. Most of the 100–150 industrial sites that line the river's banks are petrochemical plants that manufacture products from petroleum, such as organic chemicals, pesticides, gasoline, plastics, and synthetic fibers. The 13 Louisana parishes (counties) that depend on the Mississippi for drinking water have among the highest U.S. mortality rates for several forms of cancer—including rectal cancer, a disease often linked to drinking water. Among the 13 parishes, rectal cancer rates are highest among those living downstream from or within the Cancer Corridor.

Millions of pounds of unidentified chemicals buried near the plant sites threaten both surface and groundwater supplies. Nearly half of the buried drums leak. However, we need not wait for the storage drums to rust and leak. Hazardous waste can be legally injected into the ground, with the obvious potential to pollute aquifers. Although there is some regulation of these injections it is not adequate to protect our aquifers. According to the Legal Environmental Assistance Foundation, at least 25 states have documented evidence of problems caused by underground injection of hazardous waste.

Groundwater pollution is essentially permanent, because groundwater recycles slowly, remaining in aquifers for an average of 1,400 years compared to 16 days for river water. Roughly 64 percent of America's liquid hazardous waste is directly injected into the ground.[19] Nearly three times more liquid waste is injected into the ground than is poured into our rivers. Texas (oil refineries) generates one-third of America's hazardous waste; Texas and Louisiana do more than two-thirds of the injecting. In southern Louisiana, chemicals discharged by the petrochemical plants and other industrial sites are regularly found in public drinking water. Cancer rates are abnormally high within the corridor.

There are signs along many of America's rivers warning of pollution by mines, sewage, pesticide runoff from farms, or other sources. Runoff of pesticides and fertilizer from America's farms is one of the country's most vexing and unsolved pollution problems. Runoff from farms along the Mississippi River has so polluted the nearshore waters of the Gulf of Mexico that a "dead zone" has formed around the mouth of the river.[20]

### Major Sources of Pollution

Listing in rank order the major pollution producers can generate controversy. But it is generally agreed that the number one polluter in the United States is the American military. It is responsible each year for the generation of more than one-third of the nation's toxic waste[21] (Army, 37 percent; Air Force, 26 percent; Navy, 16 percent; inactive military sites, 20 percent), an amount greater than the five largest international chemical companies combined. Our domestic military installations contain more than 20,000 toxic sites and just under 10 percent of all bases are on the federal Superfund list.[22] These defense sites have contaminated an area larger than Florida. But the American public can do nothing about this appalling situation. The EPA is forbidden to investigate or sue the military. The Defense Department spends a minuscule 1.5 percent of its budget on environmental concerns, and there has been a steady decline in environmental funding in recent years.[23]

Second on the list of major polluters is the chemical industry. They produce most of the tens of thousands of organic compounds used in manufacturing and agriculture. These compounds are part of the technological society that provides our high living standard. We cannot stop using them unless we return to technologically simpler lives. The problem lies in disposal of the chemicals after use. They are legally dumped into our groundwater, our lakes and our rivers. The amount of industrial pollution dumped into our rivers, streams, and lakes rose 26 percent between 1995 and 1999.[24] The U.S. Public Interest Research Group (PIRG) released a report in 2003 reviewing state-by-state releases of toxic chemicals into our water and air in 2000, according to EPA data (table 1.2). Sort of quenches your thirst, doesn't it?

Compounding the problem is illegal dumping. An EPA report in June 2003 revealed that one-fourth of industrial facilities are not complying with their Clean Water Act discharge permits and that only one in seven of these noncomplying firms are fined for their infractions. Those who are

**Table 1.2**
Releases of toxic chemicals into the environment in 2000 and the medical problems the chemicals are known to cause

| Amount of chemicals (pounds) | Malady caused |
| --- | --- |
| >100 million | cancer |
| >138 million | birth defects, learning disabilities |
| 50 million | reproductive disorders |
| >1 billion | neurological problems |
| >1.7 billion | respiratory diseases |
| 1.6 million of lead and its compounds | neurological problems, learning disabilities, behavioral problems |
| 166,000 of mercury and its compounds | neurological problems |

*Source:* U.S. Public Interest Research Group, http://uspirg.org/uspirg.asp?id2= 8822&id3=USPIRG&.

fined pay an average of only $5,000 per incident, hardly burdensome to a major chemical company.[25]

The problem with changing industrial practices is that although injecting pollutants deep underground commonly poisons our aquifers, it is an inexpensive way to get rid of unwanted chemicals. The politically powerful chemical industry does not want to repeal federal laws that permit underground waste disposal. A better and potentially safer method of disposal is incineration, which can convert poisonous chemicals into harmless compounds. But it is very expensive. Changing from underground waste disposal to incineration will increase the cost of many products. Are we willing to pay for this?

Nearly all of America's produce is grown with massive help from pesticides, our third most serious source of pollution. These chemicals are washed by rainfall off the cropland into surface waterways and large quantities of the pesticides drain downward into our groundwater supplies. Are congressional representatives from farm states likely to vote for laws that restrict pesticide use or that require crops to be grown organically, without the use of pesticides (chapter 4)? Not likely.

A rapidly growing source of water pollution is animal excrement from the increasing number of factory farms in the United States. Americans eat a lot of meat, one million animals per hour,[26] and live animals produce a lot of urine and feces (figure 1.6). In 2000, 2.7 trillion (2,700,000,000,000) pounds of animal waste was produced by cattle (82.1 percent), hogs (12.1 percent), poultry (5.5 percent), and sheep (0.3 percent). For each pound of steak, a cow produces 53 pounds of feces and urine. Animal farms produce 86,600 pounds of excrement *every second*, more than 130 times the amount of waste that people do.[27] Texas produced twice the amount of animal excrement as the next leading state and twenty-eight times the waste of New York City's human population. Talk about being "full of it"!

Farm-animal production increased by 25 percent between 1980 and 1997. Most of this growth has occurred in large operations called *factory farms* that gain efficiency by raising animals in controlled indoor environments; manure is pumped into huge open air pits. From there it is sprayed onto agricultural fields. But the amount of waste applied often exceeds what the crops can absorb, leaving the rest to evaporate or run off into surface waters. In addition to this runoff, waste pits have cracked or leaked, killing hundreds of thousands of fish and seriously contaminating drinking water and soil. According to the Sierra Club, concentrated animal feeding operations have polluted 35,000 miles of rivers.

**Figure 1.6**
Feedlot for cattle showing normal concentrations of wastewater, feces, and urine (photo courtesy Soil Conservation Service).

## Cures

In principle the cure for most water pollution is simple. Stop pouring noxious chemicals into our waterways and stop injecting them below ground. Ending these activities will not end all water pollution because some pollution results from other sources, such as water drainage through waste piles at abandoned mines, farm-animal waste, leakage into aquifers from landfills (chapter 3), and other sources. But ending conscious and deliberate pollution will certainly bring an end to the worst offenses.

In practice the cure for most water pollution is fraught with political difficulties, which is the reason the cure has not been accomplished. Perhaps most difficult to deal with is the military. A claim of "national security" always trumps public concerns about the environment. But surely it is possible to establish a congressional committee whose members have security clearances and who could investigate pollution by the military without compromising national security.

As we will see throughout this book, the ways to end or seriously curtail environmental pollution are known. Scientists know how to solve most problems of environmental pollution or degradation. The problems are political. Are Americans concerned enough to stop pollution if it increases their cost of living? Will they vote for politicians who promise to be hard-nosed on this issue? Sooner or later they will have to. There is no other choice.

## The Quest For Perfection

Filthy water cannot be washed.
—West African proverb

For practical, political, and economic reasons, it is impossible to eliminate all forms of pollution from the environment. In the long run, it may turn out that humans are the most lethal and infectious virus on earth, able to infect and destroy all other living forms. With regard to the purity of water and air, the goal of zero contaminants, as desirable as it may be, is usually unattainable. Furthermore, even when it is obtainable using available technology, the cost of removing the last few remaining units of a noxious substance from the water or air can be astronomical. Removing the first

75 percent may be easy and cheap, but removing the next 20 percent can cost twice the amount of the initial 75 percent. The next 4 percent may cost five times as much, and the last 1 percent may require an extremely costly herculean effort. The technology may be there but at what cost-benefit ratio?

Therefore, an acceptable level of contamination needs to be determined in relation to the possible risk to human health and life. Suppose an organic contaminant in the water has been found in the laboratory to be carcinogenic when fed in measured amounts to rats, mice, or bunny rabbits. How do we assess the risk to humans of small amounts of the substance? Data from experiments on humans are lacking (no volunteers) and the experts disagree on what constitutes a safe level of exposure. Science cannot give a certain answer. Hence, economics and politics dictate what will be considered an acceptable level. In the area of water pollution, a good example is arsenic, long known to be harmful in "large" amounts. The acceptable amount was set at 50 parts per billion parts of water by the U.S. Public Health Service in 1942, which the EPA estimated in 1988 would result in a skin-cancer risk of 1 in 400 and estimated in 1992 would cause an internal cancer risk of 1.3 per 100 people. In 2002 the EPA lowered the acceptable limit to 10 ppb.[28] Other examples of limits established by other than purely scientific means include the dangers of inhaling ultrafine particles in the air and establishing a safe limit for radon exposure (chapter 7). In many cases we simply do not know what is safe (your body can deal with it without harm) and what is not.

## The Value of Human Life

But sweet, sweet is this human life,
So sweet, I fain would breathe it still.
—William Cory, *Mimnermus in Church*

How much is a human life worth? How far should the government go to save lives by reducing everyday hazards? Life is priceless, of course, especially when it is yours or a loved one's. Yet governments have budgets and must try to weigh costs and benefits. Unpleasant as it is to face the question of what someone's life is worth, the realities of the world we live in require an answer. Surely, a human life is worth at least a few hundred dollars. But is it worth a few billion?

Kip Viscusi of the Harvard Law School has researched this question.[29] What are people willing to pay to reduce the risk of death at their place of work, and how much money will they accept as compensation for an increased risk of dying on the job? The answer he determined, based on many surveys, is around $7 million. On this basis, many federal regulations fail a basic cost-benefit test (table 1.3). Only about half the regulations he studied were "cost-effective" as defined by saving a life at the cost of less than $7 million.

According to John Morrall, an official at the Office of Management and Budget twenty years ago, environmental regulations such as restrictions on different kinds of pollution generally cost over $1 billion for every life saved, and often much more. The cost of these regulations is far higher than the results seem to justify, based on what Americans believe is the value of a single life.

## Conclusion

There are many reasons to be concerned about the future adequacy of America's water supplies. Current uses are depleting or contaminating

Table 1.3
The cost of selected federal regulations

| Regulation (year issued) | Cost per life saved |
| --- | --- |
| Child-proof lighters (1993) | $100,000 |
| Respiratory protection (1998) | 100,000 |
| Logging safety rules (1998) | 100,000 |
| Electrical safety rules (1990) | 100,000 |
| Steering-column standards (1967) | 200,000 |
| Hazardous-waste disposal (1998) | 1.1 billion |
| Hazardous-waste disposal (1994) | 2.6 billion |
| Drinking-water quality (1992) | 19 billion |
| Formaldehyde exposure (1987) | 78 billion |
| Landfill restrictions (1991) | 100 billion |

*Source: The Economist,* January 24, 2004, p. 9.

many of our most important supplies, and once supplies are depleted or contaminated it may be impossible to replace or cleanse them. Even when cleansing is possible it is always very costly. As with many things in life, an ounce of prevention is worth a pound of cure.

We have made impressive gains over the past few decades in restoring and protecting our water resources, but much more remains to be done. The chief reason for the gains has been federal legislation such as the Clean Water Act. Resistance is growing, however, to the enormous investments that continually must be made in treating municipal and industrial wastes. But there is no alternative. If the way we protect our military establishment, operate our industries, and run our farms results in the pollution of our water we have no choice but to either change the way we live or treat the problem we have created.

We are often told that the two certainties in life are death and taxes. To that we can add a third: the cost of water is going to increase, most probably by a significant amount. But it is so cheap now that Americans can handle it. We will grumble and recall "the good old days" of cheap water, but we will survive. After all, if we are not willing to pay for guaranteed supplies of fresh, clean water, what are we willing to pay for?

# 2

# Floods: Too Much Water

The only thing that stops God from sending another flood is that the first one was useless.

—Nicholas Chamfort

Floods are the most commonly experienced natural calamities and have killed more people than any other natural disaster. In the United States, they account for about 60 percent of federally declared disasters and cause an average of $6 billion in damage annually. According to data kept by the National Weather Service, annual flood-related losses have increased steadily since 1930 and totaled about $100 billion in the 1990s, a fivefold increase when adjusted for inflation.[1] Increasing national wealth since 1930 accounts for most of the increase. As a 1966 task force on federal flood control policy report eloquently stated, "Floods are an act of God; flood damages result from the acts of [people]."

Similarly, annual fatalities related to floods have steadily increased from about 80 in 1930 to more than 130 today. America's population has increased by 50 percent since 1930, so there has been a slight increase in deaths from floods on a per capita basis. Floods are expensive calamities that directly affect most states and indirectly affect all Americans through their taxes. The federal government regularly subsidizes rebuilding in flood-ravaged areas through grants and low-interest loans. Where and why do floods occur? Can they be prevented or, if not, can they be controlled so that loss of life and property are lowered?

Floods can be conveniently divided into two varieties, those that originate on land and those that originate at sea. Those that originate on land can be subdivided into slow-rise floods and flash floods. Flooding that originates at sea consists of walls of ocean water carried landward, often for several miles inland, by hurricane-force winds. These various types of floods have different origins and, therefore, different characteristics. What causes these types of watery disasters and what can we do to minimize loss of life and damage to property when they occur?

### Slow-rise Floods

It wasn't raining when Noah built the ark.
—Howard Ruff

The summer of 1993 witnessed the costliest and most devastating flood in U.S. history. It killed 48 people and caused property damage totaling $16 billion in the Dakotas, Minnesota, Wisconsin, Illinois, Iowa, Nebraska, Kansas, and Missouri. In the 7 months from January through July more than a year's worth of rain fell. We are all familiar with the story of Noah's ark, when it rained for 40 days and 40 nights. Well, in Iowa it rained for 50 out of 55 days in June and July 1993. Although the entire earth was not flooded, 17,000 square miles of agricultural and urban land in the Midwest were under water for weeks. The flood crest on the Mississippi was 46 feet above normal and 3 feet above the highest ever recorded. The flood crest moved downstream for two months, from mid-June to mid-August. Forty-five recording stations had flood peaks higher than the "100-year flood" (a flood expected to occur, on average, only once every hundred years). At St. Louis, the Mississippi River was above flood stage for 144 of the 183 days between April and September. No one had ever seen Nature in such a liquid fury before.

At least 50,000 homes were damaged or destroyed and 85,000 residents had to evacuate their homes. In Des Moines, Iowa, the residents were without drinking water for 12 days. Many farm animals perished and farmers were financially devastated. Crop losses were spectacular: $1.5 billion worth of soybeans in Illinois; $1 billion worth of corn in Iowa. The failure of 388 wastewater plants spread the effects of the flood far beyond the actual flooded areas. Hazardous waste was released from 54 sites on the federal government's most-polluted list (Superfund sites) and spread over a wide area. The Federal Emergency Management Agency (FEMA) responded quickly to the disaster to supply such necessities as water-purification equipment to hospitals and to help with applications for federal aid for rebuilding costs. But Washington could not make up flood losses dollar for dollar; states, local governments, private charities, and the victims themselves had to bear much of the cost.

Why did this disaster happen? Could it have been prevented? Will it happen again? These are some of the many questions the devastated residents of the nine affected states asked. What are the answers?

### The Cause

About 33,000 feet above the earth, a distinctive meandering river of air about 100 miles wide called the jet stream flows from west to east. Its location shifts constantly, moving south during the winter and north during the summer. This distinctive river of fast-moving air exists because of interaction between the earth's rotation and the atmosphere. Typical velocities for the jet stream are 50–100 miles per hour.

Pilots, of course, are well aware of the jet stream and use this river of swiftly moving air as a tailwind to decrease flying time from, say, San Francisco to New York. On the return flight they try to avoid the jet stream. During summer months the jet stream is normally near 45° latitude and the tailwind provided by the jet stream shortens the west-to-east flying time by 30–45 minutes. During the winter the jet stream is normally over northern Mexico and the southern United States.

During the summer of 1993 the river of eastward-flowing air called the jet stream stalled over Iowa and the Midwest and formed a barrier to moisture-laden air moving northward from the Gulf of Mexico. The Gulf air bumped into the atmospheric fence formed by the jet stream and dumped its moisture without letup. Water from the surface of the Gulf of Mexico evaporated in the summer heat offshore of Texas, Louisiana, Mississippi, and Alabama, moved north, hit the jet-stream fence, dropped to the ground, and drowned nearly 20 million acres of land in the Midwest on its way back to the Gulf. This cycle continued for weeks. Clearly there is no way for humans to stop this natural, if bizarre, process. It happens every now and then for unknown reasons. We cannot control the meandering movements of the jet stream or the northward movement of Gulf air.

## Dealing with the Deluge

Never face facts; if you do you'll never get up in the morning.
—Marlo Thomas

Without human help, nature does make some preparation for flooding along river courses. We call these preparations floodplains and levees. Water in a river is normally confined to its channel, a V-shaped trench the river has sculpted in the land surface. During the mild floods that occur almost every year the water overflows the channel and spills out onto the flat ground that surrounds it. In the Midwest there is lots of flat ground so that floodplains—the area covered during floods—may be quite wide, perhaps hundreds of feet or more. At the borders of the floodplain the river constructs levees, accumulations of mud formed as the floodwaters reach their greatest extent and drop their sediment load. Levees are naturally formed ridges at floodplain margins. The floodplain-levee combination can handle later floods that are no greater than the flood that built the levee. If a larger flood occurs it will overtop or breach the levee wall and enlarge the floodplain, flooding a larger area than that of the earlier flood. So the levee farthest from the river channel marks the lateral extent of the largest flood.

When an exceptional flood is imminent, we try to help nature by piling sandbags on the top of the natural levee in an attempt to contain the floodwaters. Sometimes we are successful, sometimes not. The floodwaters may rise higher than our sandbags, or the natural levee may be breached in a place we haven't sandbagged yet. During the 1993 flood at Van Meter, Iowa, west of Des Moines, a sandbag wall built by hundreds of volunteers and National Guard troops was no match for a river that was rising nearly 1 foot per hour. The floodwater swept under the sandbags and pushed them aside on its way eastward into Des Moines. Damage in Des Moines exceeded $1 billion. Many other levees failed to hold as well. Overall, some 58 percent of the 1,400 levees on the Mississippi and Missouri Rivers were overrun or breached by the water. The U.S. Geological Survey estimated that a flood of this severity occurs there only once every 500 years, on average.

## Flash Floods

You can plan or not plan and it doesn't make a hell of a lot of difference.
What makes a difference is how much it decides to rain.
—Mark Twain, *Life on the Mississippi*

Flash floods are local events that occur very rapidly in semiarid mountainous regions. Following an unusually heavy rain (thunderstorm) in the
headwaters of a mountain stream, the water comes roaring down the
canyon in a wave often 10–15 feet high, a churning mass of water, rocks,
mud, and debris with the power to rip up trees and crush buildings in its
wake. In the United States common locations for flash floods are the
Rocky Mountain region and desert areas in New Mexico and Arizona. Because of the high velocity of the water in the narrow stream channel,
people living downstream may have little warning before the deluge hits.
Consequently, there is limited opportunity for an organized response. The
emphasis must be on saving lives rather than reducing property damage.
Even when a few hours of warning time is available, it may be impossible
to save real estate. The volume of water in a flash flood is so great that
most structures in the path of the flood cannot be protected. The only
thing to do when a flash flood is coming is get out of the way by heading
to higher ground.

## Damage Reduction

I can find no truly independent review, appraisal, or check of the economic evaluation of a federal water project which confirms the benefit-cost ratio calculated by
the construction agency.
—Herbert Marshall, *Water Research*

The existence of a floodplain around a river is nature's way of warning
us. The message is: " Keep out of here because I visit often." But humans
like to think they are above heeding warnings from nonliving things and
so they move there anyway. After all, the soil in a floodplain is very fertile and great for my crops because of its high content of nutrient-rich organic matter. And the flowing water is a great way to get rid of the liquid
waste from my factory. And I like to live next to a scenic river. It restores

my soul. Well, that's fine for your soul but probably bad for your body and possessions.

Industrial, agricultural, and urban development on floodplains has been questioned for at least 65 years. In 1937 the *Engineering News-Record* asked: "Is it sound economics to let such property be damaged year after year, to rescue and take care of the occupants, to spend millions for their local protection, when a slight shift of location would assure safety?" What should be public policy on this issue? At what point does an individual's right to live where he wants conflict with his obligation to keep his hand out of my wallet, which funds federal and state restoration efforts? As the twenty-first century progresses, American society is starting to come to grips with this question.

### Urban Sprawl and Flooding

Undeveloped land, be it soil or wetland, is a barrier to flooding. Plant roots and soil pores act like sponges, holding large amounts of water and so at least slowing the rate at which excess stream water moves away from the stream. Thus, it is no surprise that urbanization—stripping the natural vegetation, paving the land surface with impermeable cement, and erecting buildings—allows the floodwaters to spread further and faster. Floods are more frequent, and loss of life and property are greatest, in areas that have lost the most wetlands. Smaller floods are more affected by urbanization than are larger ones because large floods overwhelm the storage capacity of the soil or wetland buffer, no matter how extensive.

### Dams

One method Americans have tried is to build barriers to the downstream movement of floodwaters. We call these barriers dams. More than 75,000 dams more than 6 feet high exist in the United States (figure 2.1). On average, we have constructed one dam every day since the signing of the Declaration of Independence. Some rivers need only a single dam, whereas along others, two or three may be necessary to manage the water volume. The Mississippi, our biggest river, has 28 locks and dams, and 45 dams have been built along its tributaries. The pinnacle of dam construction was reached in 1976 with the completion of a 34-dam complex along the

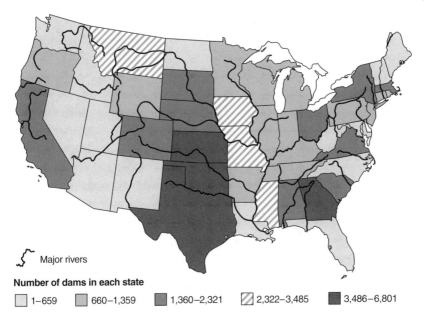

**Number of dams in each state**

☐ 1–659    ☐ 660–1,359    ■ 1,360–2,321    ▨ 2,322–3,485    ■ 3,486–6,801

**Figure 2.1**
Distribution of existing U.S. dams, 2001 (Federal Emergency Management Agency and U.S. Army Corps of Engineers).

Tennessee River and its tributaries (figure 2.2). About 75 percent of America's potential hydroelectric power has now been tapped. Of large rivers in the United States, only one—the Yellowstone—flows freely along its 600-mile length. Dams, many with heights of hundreds of feet, not only offer some control over floodwaters but turbines installed in them can generate large amounts of electricity. Hydropower generates 10 percent of America's electrical power. In addition, the reservoirs behind dams are a stable water supply during dry years and also increase tourism in the newly scenic area.

But many dams are old and should be retired for safety reasons. The average life expectancy of a dam is 50 years, and one-quarter of all America's dams are now more than 50 years old. By 2020 that figure will reach 85 percent.[2]

But dams have their drawbacks, and these have only recently made their way into public consciousness.

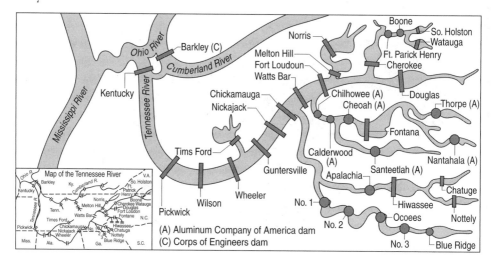

**Figure 2.2**
Diagram of the TVA water-control system built during the nation's dam-building frenzy (Tennessee Valley Authority).

1. The reservoirs behind them fill up. As the stream behind them stops it drops its load of sediment, mud and sand. Sediment fills between 0.5 and 1 percent of reservoir capacity each year. Many existing reservoirs are now useless, filled with mud. Recent data suggest that 20 percent of America's reservoirs will function for less than 50 years and only half will function for 100 years.

2. Dams may collapse. The Association of State Dam Safety Officials estimates that about 30 percent of America's 75,000 dams are hazardous. Some dams are poorly constructed, some were built in locations likely to be hit by earthquakes, some fail because of overtopping by large floods, some fail because of landslides from surrounding steep slopes, and others are simply old and have become structurally weak.

One catastrophic example of dam failure and resulting catastrophic flood occurred at the Teton Dam in Idaho, a 300-foot-high earthen dam with 250,000 acre-feet of stored water. On June 5, 1976, water leakage through the dam weakened the structure and the dam failed, killing 11 people, making 25,000 homeless, and inflicting about $400 million in damages downstream to the Teton–Snake River Valley. When a large dam fails, which often happens suddenly, it may be impossible for those downstream to get out of the way.

3. Environmentalists have emphasized the ecological damage caused by dam construction and point out that the true cost of a dam never shows up on a balance sheet. Dams and their reservoirs drown large amounts of property, displace farms and people, drive out wildlife, kill vegetation, cause erosion of coastal deltas, and alter ecosystems. Seventeen percent of the country's river miles are buried beneath reservoirs. As is normal in ecological controversies, the amount of ecological damage and its significance are difficult to calculate and are therefore disputed.

Because of these negative factors associated with dams, few are being constructed these days in the United States. And in 1996, for the first time, a dam was removed for purely environmental reasons, to restore salmon, sturgeon, and other migratory fish populations. Dams are coming down from Maine to California. By the end of 2000 more than 500 dams had been dismantled, most of them among the 82 percent of American dams that have less than 100 acre-feet of reservoir storage. It is unclear how far the present dam-busting frenzy will go but both proponents and opponents generally agree that the era of large dams in the United States is ending, although some new high dams are still coming online. In January 2000, the Seven Oaks Dam, the twelfth-highest dam in the world, was dedicated in southern California. Its purpose is to lessen the effects of catastrophic floods in a densely populated area east of Los Angeles.

### Artificial Levees

Natural levees are usually low. By constructing higher levees along the channel margins, the channel can hold more water, reducing the occurrence of floods. The federal government has advocated the construction of artificial levees for more than a century. Over the years the U.S. Army Corps of Engineers has built levees along the Mississippi River and its tributaries that now total 7,000 miles in length, 2,500 miles of them on the Mississippi itself. Some of these levees are 30 feet high. However, the practice of levee building is now being reassessed.

Many of the artificial levees along the Mississippi River could not contain the 1993 deluge and much land was flooded that had long been protected. Calculations indicate that if the levees had been high enough to contain the 1993 floodwaters, the river would have risen about 6 feet higher downstream at St. Louis, putting much of that city under water. The waters were barely contained by the existing 50-foot levee wall that protects the population of the city. Only 3 feet of artificial levee separated the city and 15 million acres of farmland from disaster. If the water doesn't spill out upstream, obviously there will be more water in the channel-floodplain system downstream. Salvation upstream means condemnation downstream.

The legacy of levee construction and straightening the river by cutting channels across large meander loops has had effects downstream of St. Louis, effects only now being fully recognized. The increased velocity of Mississippi River water in its lower reaches now carries to the Gulf of Mexico mud that once settled on the delta floodplain on which New Orleans sits. As a result of this sediment deprivation, the delta is retreating more rapidly than it otherwise would, retreating by more than 2 inches per day in some areas. Much of New Orleans is already below sea level and the levees keeping the river at bay have to be periodically raised. Restraining the Gulf of Mexico is more difficult.

### Floodplain Zoning

The growing lack of interest in new dam construction and the demolishing of some existing ones pose increased problems for cities located along rivers. These problems and concerns are increased by the current questions about the long-range desirability of artificial levees. It is agreed that large cities must be protected. No one favors tearing down the levee that protected St. Louis in 1993. But there is debate about the best way to protect life and property in the light of present knowledge and public sentiment.

Floodplain zoning is one good option, but the idea has received a less-than-enthusiastic response from local governments. There is intense local political pressure in many parts of the country against restrictions on development, particularly from real estate interests. Real estate developers almost always are opposed to restrictions on the uses (residential, commercial, industrial) to which they may put their property. Americans in general do not favor restrictions on the use of private property, and most floodplain land is privately owned. Zoning laws are often accompanied by building codes or construction standards that include floodproofing requirements such as placing shields around buildings, or erecting buildings on stilts that raise the bottom floor to several feet above ground level.

Perhaps it would be more acceptable to simply require that prospective buyers or builders be informed by the local government of the risk they would be taking by building on the floodplain. If they then chose to take this risk, at least they would be aware of the danger. In 1997 California became the first state to pass such a law, which took effect in June 1998. The law requires that a seller give a prospective buyer a "Natural Hazard Disclosure Statement" that includes information on the potential for flooding, whether there is a fire hazard in the vicinity, and the location of the property in relation to potential earthquakes.

Even in areas where land use is controlled, there may still be the potential for loss of life and damage from very large floods. There are no iron-clad guarantees of safety near riverbanks. However, a carefully prepared plan should be developed defining who does what in case of an emergency. This involves the police, civil authorities, and all emergency services and requires a clear understanding of what might happen, where, and how quickly. It therefore relies on accurate risk assessment to specify the areas likely to be flooded and the escape routes that might be used. Emergency response plans should be developed and reviewed from time to time in consultation with all concerned and should be clearly announced to the public.

In an effort to encourage local communities to stop destroying wetlands and building in floodplains, to manage urbanization wisely, and to include flood prevention in their land-use planning, the Federal Emergency Management Association (FEMA) inaugurated a program called Project Impact: Building a Disaster Resistant Community. FEMA helps governments work with businesses, educational leaders, and environmentalists in their communities. Together they can alter zoning laws, buy out floodplains, and discourage potentially disastrous development.

### Relocation
At least one town decided it had all it could take from floods.[3] The people of Valmeyer, Illinois, were inundated twice in the summer of 1993 by the rampaging Mississippi River and saw 90 percent of their homes, offices,

and public buildings destroyed. They decided in October to move the entire community to a site 300 feet above the Mississippi floodplain. Following scientific examination of the site by the Illinois State Geological Survey, the 900 citizens voted in December 1993 to apply for relocation, which was approved in May 1994.

Valmeyer's relocation cost $9.6 million, of which the federal government supplied $7.2 million. The remaining $2.4 million, divided among 900 residents, came to less than $3,000 per person. The cost for a family of four was about $11,000. As a condition of the grant, the town was required to adopt drainage, sediment, and erosion controls; a storm-water management plan; construction guidelines; and other ordinances to avoid activating known geological hazards in the area.

The citizens of Valmeyer believed it was in their best interest to adapt to their surroundings rather than continue to fight what was surely a losing battle. They showed uncommon good sense in their decision to relocate. Unfortunately, their example has not been followed by other small communities threatened by flooding. And, of course, large cities do not have the option exercised by Valmeyer.

## Hurricanes

I hunger for the sea's edge, the limits of the land.
Where the wild old Atlantic is shouting on the sand.
—John Masefield, *A Wanderer's Song*

Hurricanes cause extensive flooding, loss of life, and property damage in coastal areas. They form over tropical ocean waters where winds are light, the humidity is high in a thick layer, and the sea-surface temperature is at least 73°F; the peak in hurricane generation is reached at 84°F. These conditions provide what a hurricane needs most, plentiful moisture and heat energy. A rise in sea-surface temperature of 1°F causes a doubling in the number of hurricanes.[4] Hurricanes are rated in categories from 1 to 5 based on wind speed, which can reach 200 miles per hour (table 2.1). In an average year about five hurricanes develop that might threaten the Atlantic and Gulf coasts.

**Table 2.1**
Categories of hurricane intensity, the Saffir-Simpson Damage Potential scale

| Category | Magnitude | Wind speed (mph) | Storm surge (ft) | Damage |
|---|---|---|---|---|
| 1 | mild | 74–95 | 4–5 | Minimal; damage mainly to trees, shrubbery, and unanchored mobile homes. Examples: Gustav, 2002; Claudette, 2003. |
| 2 | moderate | 96–110 | 6–8 | Moderate: some trees blown down; major damages to exposed mobile homes. Examples: Isabel, 2003; Irene, 1999. |
| 3 | severe | 111–130 | 9–12 | Extreme: foliage removed from trees; large trees blown down; mobile homes destroyed; some structural damage to small buildings. Examples: Isadore, 2002; Fabian, 2003. |
| 4 | very severe | 131–155 | 13–18 | Extreme: all signs blown down; extensive damage to roofs, windows, and doors; complete destruction of mobile homes; flooding inland as far as 6 miles; major damage to lower floors of structures near shore. Examples: Floyd, 1999; Isabel, 2003. |
| 5 | catastrophic | greater than 155 | greater than 18 | Catastrophic: severe damage to windows and doors; extensive damage to roofs of homes and industrial buildings; small buildings overturned and blown away; major damage to lower floors of all structures less than 15 feet above sea level within 1,500 feet of shore. Examples: Gilbert, 1988; Mitch, 1998. |

When a hurricane makes landfall the intense rainfall, storm surge, and resulting flooding can be almost unimaginably devastating. One day's rainfall from a moderate hurricane equals the average yearly discharge of the Colorado River at its point of greatest flow. But as fearsome as the rain is, the storm surge is most deadly. The storm surge is a rise in sea level (a hill of water) along the shore, caused by a drop in atmospheric pressure inside the hurricane, which allows water from surrounding higher-pressure areas to push in, creating a bulge. As this happens, the wind piles up water against the shore. The height of a surge can exceed 40 feet, a volume of water that can sweep away most human-made construction.

Modern weather satellites track hurricanes from their inception in the tropical Atlantic to their eventual loss of intensity as they move inland after making landfall. The advance warning from the National Hurricane Center (figure 2.3) means there are few deaths from the floods they cause, even in the worst storms. But financial loss is another story. Although the number of deaths has decreased over the years, dollar losses have increased considerably because Americans are clustering in ever-increasing numbers near the coast and are building ever-more costly structures. Damages from individual hurricanes regularly total billions of dollars, the costliest so far being the $26.5 billion in damages wrought by Andrew in 1992.

There are about 45 million permanent residents along the hurricane-prone coastline and the population is still growing. Florida, where hurricanes are most frequent, leads the nation in new residents. In addition to the permanent residents, the holiday, weekend, and vacation populations swell in some coastal areas 10- to 100-fold.

In greatest danger are structures located on the many sandbars and barrier islands that fringe the mid-Atlantic Coast, particularly the Outer Banks of North Carolina. These linear accumulations of sand are separated from the mainland and are only a few feet above high-tide level. They are easily drowned by storm surges. Nevertheless, expensive beach-front homes proliferate on these unstable sand accumulations and are regularly rebuilt after the latest hurricane has passed. How can the owners afford this luxury? The answer is that reconstruction of their homes is subsidized by the federal government.

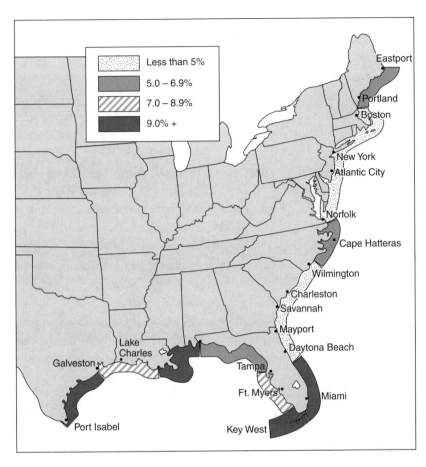

**Figure 2.3**
Probability of hurricane landfall along the Atlantic and Gulf coasts (National Oceanographic and Atmospheric Administration).

Insurance

Oh, dry the starting tear, for they were heavily insured.
—William Schwenck Gilbert, *Etiquette*

When natural disasters such as floods occur, federal disaster-relief funds are often a part of the rescue, recovery, and rebuilding operations. The nation's taxpayers pay part of the damage suffered by those who have been living in high-risk areas, be the area a river floodplain or an offshore sandbar in a hurricane-prone area. The National Flood Insurance Act of 1968 was enacted as a subsidy for those living in such threatened areas. It was established because by the 1960s private insurers had largely opted out of the flood-insurance business because of the increasing risk from unwise, flood-prone development. Under the federal program, homes can now be insured for up to $250,000; furnishings and belongings for up to $100,000. The average annual premium is only $320.[5]

From 1978 to 1995, nearly 40 percent of federal flood-insurance payments went to about 2 percent of properties that repeatedly flood.[6] Among them were about 5,600 properties whose owners received cumulative payments in excess of what their homes were worth. In Houston, a single house valued at $114, 480 was flooded 16 times from 1989 to 1995. The owners collected payments totaling $806,590, about seven times the value of their home.[7] Similar tales come from North Carolina, Louisiana, and California. Given that modern advance-warning systems minimize the danger of loss of life in flood disasters, it almost pays to build in flood-prone areas.

Looking back over disaster declarations and expenditures for the years 1965 to 1997, researchers at the National Center for Atmospheric Research found that presidents were 46 percent more likely to declare flood-related disasters during presidential campaign years, whether or not the flooding and rains were severe.[8] Politicians are unlikely to vote to cut off funds for their constituents who have suffered natural disasters, even in nonelection years. And flooding is a problem in more than half the states, many of them with large congressional delegations. Nevertheless, it is pretty hard to justify locating in a floodplain as a subsidized public policy. Because of this, the federal government has bought 17,000 properties from people living in floodplains, and this approach may be expanded as a money-saving substitute for building ever-more dams and levees. To what extent should individuals be held responsible for making sound, informed decisions about where to live?

Private insurers are in a near-panic mode about an apparent sharp increase in property losses over the past 20 years. Since 1980 the United States has suffered 37 "billion-dollar storms," 31 of them since 1988. Losses from floods and hurricanes increased from an average of $2 billion a year in the 1980s to $12 billion in the 1990s. Some companies now refuse to issue hurricane coverage in Florida.

### Floods and Global Warming

What do you do when the past is no longer a guide to the future?
—Jesse Ansubel, *Fortune magazine*

Global warming, whatever its cause, is now a recognized fact of life to most scientists who have studied the question. And global warming has implications for precipitation. There is some indication, but not yet proof, that highly volatile weather patterns are one of the consequences of higher global temperatures. According to the National Climatic Data Center, rainfall amounts have increased in most areas of the United States and since 1990 the area of the country affected by extreme precipitation events has jumped about 20 percent. The proportion of precipitation falling in extreme events—more than 2 inches of rain in 24 hours—has risen, while the proportion coming in gentle showers—less than 1 inch in 24 hours—has declined. Intense rainfall increases runoff from the land surface, which increases the amount of water in streams and, hence, flood potential. Canada, England, and Mexico show similar precipitation trends.

The effect of global warming on the frequency or intensity of hurricanes is straightforward. With higher ocean temperatures the area of ocean warm enough to spawn hurricanes increases. But a trend in frequency or intensity of hurricanes with time is very hard to detect because hurricanes show an enormous natural variability. A trend would have to be very pronounced to be detected above the natural variability.

As an illustration of the interpretive problem, we can note that the total of 41 hurricanes during the period from 1994 to 1999 was the highest for any 5-year period since record keeping began in 1886. In addition, the 1999 North Atlantic hurricane season was the fiercest on record. Five storms reached category 4. Is this random variation or part of a trend? Floridians hope it is only random variation. Their homeowner insurance premiums have almost doubled in the past 10 years.

## Conclusion

The phenomenon of a river overflowing its banks is a common feature of streams. Riverine floods are natural occurrences that are best dealt with by getting out of their way. We should not build in flood-prone areas. We must learn to live with natural phenomena rather than try to fight them. Dams and artificial levees are not acceptable substitutes for good judgment about where to build. Natural forces such as water moving downslope must be allowed to "do their thing." Although we cannot change the present location of large cities, continued construction in floodplains should be discouraged.

Hurricanes are a permanent feature of the interaction between the atmosphere and ocean. They can be counted on to appear every year, but their number and intensity vary from year to year and cannot be predicted. Similarly, the place where a particular hurricane will hit the coast of the United States (landfall) is unpredictable. Unnecessary construction should be discouraged along the low-lying coastlines from Virginia to south Texas.

Global warming is likely to increase the frequency of floods and hurricanes, but data at present are not good enough to say that an increase has begun.

# 3

## Garbage: The Smelly Mountain

We no longer use the term *garbage*. We prefer to call it matured waste.
—Manuel Mindanao, real estate developer of housing on a ten-story pile of trash

"Did you know that the average person produces three pounds of garbage a day?" a woman asks her therapist at the start of the 1989 movie *Sex, Lies, and Videotape*. "I'd really like to know where it's all going to go." In this chapter we will explore this question and suggest some answers. We will deal with a subject most of us would rather not think about, much less be around—garbage. Officially, garbage is called municipal solid waste, because its disposal is a service we expect from municipal governments, and it is solid as opposed to liquid sewage or wastewater. The amount of solid waste each American generated each day increased by 70 percent between 1960 and 2000, from 2.7 pounds per person to 4.5 pounds per person.[1] The population grew as well during these years, so the total amount of garbage we generated grew from 88.1 million tons annually to 231.9 million tons. That's a lot of solid waste to get rid of. Americans generate about twice the amount per capita generated by other industrialized countries such as Germany, France, or Japan. America is indeed a wasteful society.

Cities and towns are responsible for only 5 percent of the solid waste Americans produce, but municipal garbage is the refuse we are most conscious of. Most of us do not see daily the solid waste produced by farm animals (39 percent), the mineral extraction and processing industries (38 percent), crop residues (14 percent), or industry (3 percent). But we cannot escape the presence and pressure of household garbage. Is there a family dwelling in the United States that has not heard the cry, "Don't forget to take out the garbage"?

So someone grumbles and takes it outside to a large metal or plastic pail, or perhaps to a dumpster on the corner in communities where house-to-house pickup has been dropped in favor of this cheaper and more efficient method of collection. Then, if there is not a strike in progress, one of America's 136,000 ancient (average age 7 years) gas-guzzling (2.8 mpg) city garbage trucks with a Mississippi-riverboat rotating-paddle-wheel gadget in the back will come to collect the trash.[2] We all have been part of this house-to-pail or house-to-dumpster experience. But what happens after that? Where does the city truck go from the last garbage-pickup point?

**What's In It?**

We are hooked like junkies, dependent on the drug of wasteful consumption.
—Raymond F. Dasmann

Before we answer this question, let's ask ourselves what we dumped in the pail or dumpster. When we think deeply about the content of garbage (not too often for most of us), things like chicken bones, pork fat, brown lettuce, or moldy broccoli come to mind. In fact, however, food residue forms only 11 percent of our solid waste (figure 3.1), thanks in large part to the presence of 45 million garbage-disposal units under the kitchen sinks in American homes. If you take a day's vacation trip to a local landfill you will see there are almost no flies there, a sure sign that there is little to eat in the vicinity. Most municipal solid waste consists of manufactured items that are essential in a modern society—paper, plastics, metals, and glass. These total 61.4 percent of urban solid waste. Over the past 15 years the biggest increase has been in plastics, up from 8.3 percent in 1986 to the present 10.7 percent, an increase of 29 percent.

One item in landfills that is increasing rapidly in abundance is old computer monitors and circuit boards. This is worrying not only because of the space a monitor takes up in a landfill but also because the average cathode-ray tube contains 5–8 pounds of lead, which can seep into the groundwater under landfills or, if the tube is incinerated, pollute the air. Printed circuit boards can also contain toxic chemicals. An estimated 315

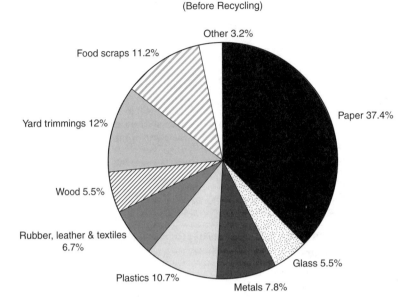

**Figure 3.1**
Types and amounts of municipal solid waste (Environmental Protection Agency, 2002).

million computers that were obsolete by 2004 contain an estimated 1.2 billion pounds of lead. In 2000, Massachusetts became the first state to ban the disposal by private individuals of computer screens in landfills and incinerators.[3]

Another rapidly increasing item in municipal solid waste is disposable diapers, first introduced to parents in 1968. As the population grows so does the number of diapers being used, now about 20 billion, and 90 percent of America's parents use only disposable diapers for baby maintenance and consider them the most essential modern convenience. In the diaper wars, cloth has folded. Reusable cloth diapers are a thing of the past. How can they compete with available colored diapers, Sesame Street diapers, and so on? From the moment a child is born until that glorious day she or he becomes potty trained, a parent will use between 5,000 and 8,000 diapers, costing about $2,000.[4] This gives new meaning to the expression "the bottom line."

## What Should We Do With It?

It is always possible to find somewhere a kind word for rubbish.
—Louis Nizer

Okay, so now the garbage truck has left its last pickup point for the day. Where does it go? Well, there are three realistic solutions for disposing of the city's municipal solid waste. We can bury it (landfills), recycle it, or burn it (incineration). At present, recycling and incineration are increasingly favored at the expense of landfills (figure 3.2). Each method has its positive and negative sides.

### Landfills

In 1976 Congress passed a law prohibiting the open refuse dumps so common in previous years, mandating instead a disposal system called a *sanitary landfill*. Tens of thousands of sanitary landfills appeared almost overnight but as of 2000 there were less than 2,000 operating landfills re-

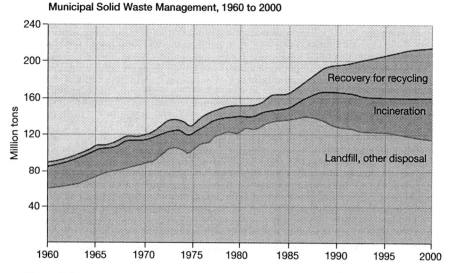

**Municipal Solid Waste Management, 1960 to 2000**

**Figure 3.2**
Disposal of municipal solid waste. Landfilling peaked in the late 1980s and is now in decline as more waste is incinerated and recycled (Environmental Protection Agency).

maining (figure 3.3). However, total landfill capacity has remained relatively constant for the past 15 years because new landfills are much larger than in the past. Landfills receive about half of our solid waste and, in some areas, have increased the size of the community (figure 3.4).

A well-designed sanitary landfill differs drastically from an open dump. The trash is placed in a hole underlain by impermeable clay and lined with chemically resistant plastic, compacted daily by heavy equipment such as bulldozers, and covered each day with a layer of dirt. Immediately above the basal plastic liner is a system of pipes and sealed pumps designed to collect rainwater and other liquids draining through the waste pile. The clay substrate, plastic liner, and pumps keep landfill liquids from descending into aquifers. The daily covering of dirt reduces accessibility to vermin, lessens the risk of fire, and decreases odor.

Unfortunately, most sanitary landfills opened decades ago, when environmental concerns were not focused as they are today. Thus, the average landfill does not meet current legal standards. For example, only about one-third of the nation's landfills have plastic liners, many lack pipes and pumps at their base, and many landfills are not underlain by impermeable clay. As a consequence, aquifers below many landfills have been polluted by liquid draining downward through the pile of trash.

**Leachate**    The waste pile that is the landfill is permeable. As rainwater percolates through the pile it dissolves many substances, just as water passing through coffee grounds dissolves chemicals in the coffee. The chemically enriched water in the landfill is called leachate. If the only things in the landfill were chicken bones and plastic milk cartons, problems would be minimal. Unfortunately, however, along with these relatively harmless items are known hazardous materials. Common examples include batteries, plywood, drain cleaners, furniture polish, bug sprays, oven cleaner, weed killer, old paintbrushes, and turpentine. These discarded items contain an assortment of toxic materials such as arsenic, lead, mercury, and complex organic compounds known as dioxins and furans. Ingesting these substances can cause congenital birth defects, disorders of the nervous system, and cancer. Pregnant women living within 1.8 miles of a landfill site have been found to have a 33 percent higher risk of having a baby with a congenital defect than women living farther away.

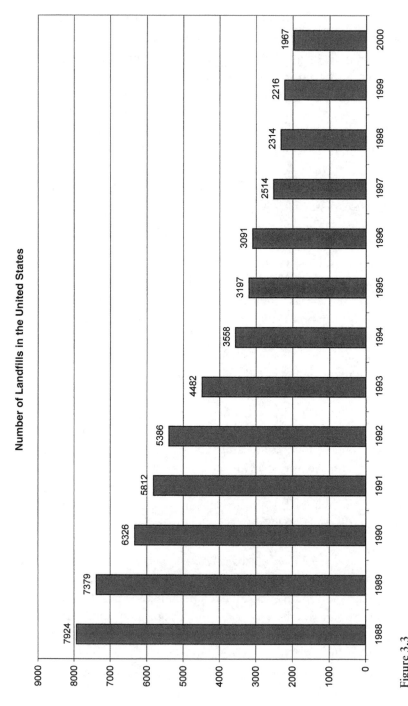

**Figure 3.3**
Landfills, fewer in number but larger in size (Environmental Protection Agency).

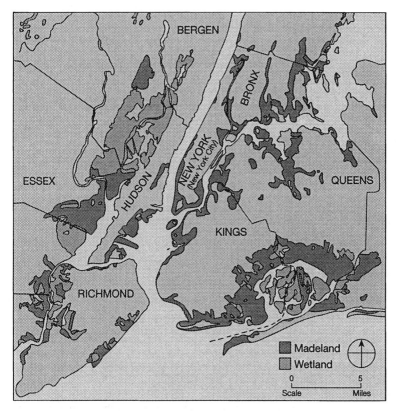

**Figure 3.4**
New York City and suburbs have grown appreciably by the addition of "made-land" created by landfilling. Much of the new land is former shallow wetland (Waste Management, Regional Plan Association, New York, 1968).

As the list of discards shows, liquids as well as solids are tossed out of American homes. Each year, every American household sends an average of 1.5 quarts of hazardous liquid to landfills. Multiply this number by 75 million households to grasp the scale of the hazardous throwaways.

Experts in landfill design believe that no landfill can be guaranteed to be permanently secure. Plastic liners can be punctured by sharp stones; vermin burrow through the waste pile and tear the basal plastic liner; acids in leachate may eventually breach the liner; the clay layer beneath the liner may contain permeable sand lenses. Eventually, all pipes clog and all buried pumps fail. Nothing built by humans is permanent. Surveys reveal that 82 percent of landfills leak and half of these have leak areas of more than 1 square foot.

**Gassy Landfills**    When organic materials such as paper, wood, or turkey bones decompose in contact with the oxygen in the air, they turn into carbon dioxide and water. In the innards of a landfill, however, there is little or no oxygen. Organic materials decompose anaerobically (in the absence of oxygen) to produce methane, an oxygenless gas. Methane produced at more than 320 landfills is collected by municipalities and used as fuel to generate electricity. Two hundred more collection facilities are either under construction or in the planning stages and about 500 additional landfills have commercial methane potential. At the world's largest landfill (closed in 2001), on Staten Island, New York, enough methane is recovered to satisfy the annual energy needs of 12,000 households. In small landfills, too little methane is produced for it to be recovered economically, so the gas is either burned off or vented into the air through perforated pipes inserted into the landfill. Accumulations of methane are potentially explosive, so venting of large landfills is essential.

In 1991, potentially explosive methane concentrations were discovered in 33 of 44 homes in a Savannah, Georgia, subdivision. Half of the affected homes were located directly above an abandoned landfill. The gas entered the homes through cracks in basement flooring. Methane is explosive at concentrations of more than 5 percent in room air. Lawsuits were filed by homeowners and the city purchased most of the homes at prevailing value. Some homeowners decided to remain because the city offered to place meters and vents in contaminated parts of the structure to preclude the possibility of an explosion.

In Japan, methane is being vented from a golf course built over a landfill and golfers who smoke are in danger of blowing themselves up. "No Smoking" signs have been placed near some holes.

**Landfill Subsidence**    Landfills are composed of irregularly shaped objects that do not fit together like a picture puzzle. The pieces compact irregularly under the weight of overlying debris. The graded flat top of an abandoned landfill will not remain flat. Settlement rate may be several inches or even feet during the first few years but decreases with time as compaction lessens. Only small structures or playgrounds can be safely built over landfills.

**Landfulls** The ultimate problem with landfills is that they become land-fulls. They fill up. About half the states and half the country's cities have no more landfill space. The decrease in landfill space has generated an enormous interstate traffic in garbage (figure 3.5). Many of America's large and populous cities are desperate. Philadelphia no longer has an active landfill and is forced to ship its municipal solid waste many hundreds of miles.

The cost of this transport is substantial. The cost of refuse collection and disposal has risen much faster than the consumer price index and faster than the cost of other utility services such as water, electricity, piped gas, telephone service, and cable television. Because of rising haulage

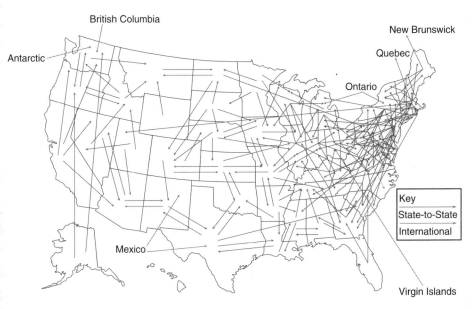

**Figure 3.5**
Simplified map of interstate traffic in garbage. Trash-shipping routes from New York and New Jersey, the biggest garbage exporters, are not shown because the map would be illegible. New York's garbage is trucked as far away as New Mexico. *Source:* Edward W. Repa, "Interstate Movement of Municipal Solid Waste," *NSWMA Research Bulletin* 03-01, May 2003. Reprinted by permission of the author.

prices and increasing population, the cost of garbage collection and disposal to the city of Philadelphia is ten times higher today than it was in 1980. Because of the ballooning cost of solid-waste disposal, some 2,000 municipalities around the country have decided to charge for waste collection by the bag rather than by flat monthly fees. Homeowners know the cost of each bag of trash and can count their savings when trash output is lowered. In these 2,000 municipalities household solid waste has been reduced by 25 to 45 percent. Getting hit in the pocketbook nearly always produces the desired result. As some wag has said, money is not everything but it certainly seems to be way ahead of whatever is in second place.

Large cities commonly pay disposal fees of more than $50 per ton to haulage contractors. There is a lot of trash and a lot of money in it. New York and New Jersey, the states with the largest trash exports, ship their solid waste as far away as New Mexico. Some states make a lot of money by accepting the solid waste from other states. Pennsylvania has increased its trash imports every year since 1991[5] and takes in about one-third of all the trash exported by states. The state increased its landfill capacity and accepted 12.6 million tons of waste in 2000, enriching the state's coffers by hundreds of millions of dollars. Federal, state, and local governments spend billions each year on waste disposal. It accounts for the majority of environmental expenditures.

As you can imagine, citizens' groups in many states that are vocal in their objection to their state being known as one of the nation's major garbage pits. Michigan's governor (2003) announced that "Our state cannot—and will not—be the nation's trash can." Michigan in 2003 surpassed Virginia as the nation's number two recipient of garbage from elsewhere. The governor's attitude is known as NIMBY—Not In My Back Yard. Within recent years new but related acronyms have been devised (figure 3.6).

At least one city is trying to recoup some of the money it spends on garbage collection and disposal. New York, by far the largest waste exporter, is reaping a financial reward from the scenic view afforded by its ownership of the world's largest landfill. The Fresh Kills landfill in New York City received garbage from 1948 to 2001 and now rises more than 150 feet above sea level and occupies 2,200 acres. Its volume is nearly 3

```
A Wall Poster Seen at EPA*

The ABCs of Waste Disposal

NIMBY...Not In My Back Yard
NIMFYE...Not In My Front Yard Either
PIITBY...Put It In Their Back Yard
NIMEY...Not In My Election Year
NIMTOO...Not In My Term Of Office
LULU...Locally Unavailable Land Use
NOPE...Not On Planet Earth

*Environmental Protection Agency
```

Figure 3.6
Acronyms unlimited in the waste-disposal trade.

billion cubic feet, larger than the Great Wall of China. Fresh Kills was decommissioned at a cost of $1 billion. The site is being turned into a nature preserve and park. New York City generates 11,000 tons of household garbage every day,[6] enough to fill a convoy of 600 trucks that stretches for 6 miles. Too much for *my* backyard!

New York City's Department of Sanitation has produced an attractive color brochure and video to help visitors find their way around. No one wants to be lost amidst a mountainous pile of garbage covering 4.7 square miles. About 300 people per week visit Fresh Kills to take a 90-minute guided tour. According to the department's assistant commissioner for public affairs, "It's a very sensory experience visually. As far as the eye can see is garbage; aurally, yes, with all that heavy equipment, and in terms of your olfactory senses, you'll certainly know where you're standing. It is one of those things in life you have to experience to truly appreciate."[7]

**Biodegradability**    Many products sold today try to capitalize on the public's increasing environmental awareness by touting themselves as biodegradable. What does this mean? A biodegradable substance is one that becomes "biologically degraded," meaning that it is decomposed by microorganisms such as bacteria and fungi, presumably within a few years.

It is always assumed that the product to be degraded is left in contact with sunlight, rain, and oxygen—that is, left on the ground. The problem, of course, is that inside a landfill there is no sunlight, and if the landfill is well designed rainwater and air are likewise kept out. Under these conditions decomposition is very slow, perhaps taking longer than a human life span. Paper forms 40 percent of the material in an average landfill, and many newspapers decades old have been excavated from landfills by garbologists. They are as readable today as when they were buried. Food in the landfill such as hot dogs is likewise intact, although there is some question about its edibility. Plastic is virtually immortal. If the pilgrims had had six-packs, the six-pack rings would still be around today. Today's landfill is tomorrow's time capsule.

Another virtually indestructible item is the automobile tire. Many landfills won't accept them because they trap methane gas generated by other garbage, breed mosquitoes in the water they trap in their concavity, and burn uncontrollably when they catch fire. Forty-four states restrict scrap-tire disposal at landfills. Americans discard about 270 million tires every year. However, 76 percent of scrap tires are used for commercial purposes. An increasing number are now burned as fuel for power plants and cement kilns. Tires provide 25 percent more energy per pound than coal and can be burned at temperatures high enough to destroy most organic pollutants. The ash can be mixed with cement or sold to fertilizer manufacturers if the toxic metals from the tires' belts are removed.

The present world's record for indestructibility is used chewing gum, found in a Swedish swamp and dated at 6,500 years.[8] Apparently the swamp was used as an unsupervised dump, because there were no landfills at the time. No one has been permitted to chew the birch-bark gum so we don't know how it tastes or, indeed, if it had some flavor when discarded. The size of the tooth marks on the gum suggests it was chewed by preteens and teenagers.

### Recycling

We all recycle. Families with children use hand-me-down clothing primarily to save money, but they are recycling nonetheless. Old cars are crushed and the metal in them recovered for use elsewhere. In 1989 California passed a recycling law that required communities in the state to cut by 50 percent the amount of trash they sent to landfills by 1999. Attainment reached 37 percent by 1999.⁹ Many states have "bottle laws" requiring deposits on beverage containers to encourage their return. And many states and communities have successful voluntary recycling programs for paper, aluminum, or other materials. As shown in table 3.1, recovery rates are highest for nonferrous (not iron or steel) metals and lowest for food (I should hope so!) plastics, and wood. Americans do a lot of recycling.

The benefits of recycling rather than throwing the item away are unarguable (table 3.2).

• Recycling 1 ton of steel prevents the production of 200 pounds of air pollutants, 100 pounds of water pollutants, and almost 3 tons of mining waste. Recycling steel cans saves 74 percent of the energy that would be used to produce them from virgin materials.

• Recycling 1 ton of aluminum eliminates the need for 4 tons of aluminum ore and almost a ton of petroleum products to provide the energy for mining. Recycling aluminum scrap saves 95 percent of the energy that would

**Table 3.1**
Recovery rates for materials in the U.S. municipal solid-waste stream

| Material | Percent recycled from municipal solid waste |
| --- | --- |
| Paper and paperboard | 41.9 |
| Metals: total | 35.2 |
| Steel | 33.6 |
| Aluminum | 27.8 |
| Other nonferrous metals | 66.9 |
| Glass | 23.4 |
| Textiles | 12.9 |
| Rubber and leather | 12.7 |
| Wood | 5.9 |
| Plastics | 5.6 |
| Other materials | 21.4 |
| Other wastes: total | 23.3 |
| Yard trimmings | 45.3 |
| Food | 2.2 |

*Source:* U.S. Geological Survey Circular 1221, 2002, p. 18.

Table 3.2
Environmental benefits of recycling

| Environmental benefit | Aluminum (%) | Steel (%) | Paper (%) | Glass (%) |
|---|---|---|---|---|
| *Reduction of:* | | | | |
| Energy use | 90–97 | 47–74 | 23–74 | 4–32 |
| Air pollution | 95 | 85 | 74 | 20 |
| Water pollution | 97 | 76 | 35 | — |
| Mining wastes | — | 97 | — | 80 |
| Water use | — | 40 | 58 | 50 |

*Source:* R. C. Letcher and M. T. Sheil, "Source Separation and Citizens' Recycling," in W. D. Robinson, ed., *The Solid Waste Handbook* (New York, Wiley, 1986, p. 220).

have been required to make new aluminum from ore. The United States throws away enough aluminum to replace its entire commercial aircraft fleet every 3 months.

• The paper industry is the largest single user of fuel oil in the United States, and every ton of paper recycled keeps 60 pounds of pollutants out of the air. Recycling a ton of paper saves about 17 trees, which live to absorb 250 pounds of global-warming carbon dioxide gas annually. Forty-nine percent of paper is recycled.

The United States had an overall recycling rate of 32 percent in 2001, up from 10 percent in 1980. Some communities have achieved 50 percent. The chief problem with recycling is its cost. It costs about $200 per ton to pick up and sort the recyclables that most communities include in their curbside programs, but the company receives only half that amount when it sells the materials—glass, aluminum, plastic, steel cans, and newspapers. The communities must make up the difference, plus a reasonable profit for the operator. As a result some communities have abandoned their recycling programs. No doubt as the cost of landfill space continues to climb, recycling will become more cost-effective.

Part of the public's enchantment with recycling is that it requires no change in the American culture of consumerism. It helps soothe the conscience of a throwaway society. Recycling is also an easy way for governments at all levels to score environmental brownie (or greenie) points with their constituents. In the environmentalists' trinity, the three Rs of solid waste—reduce, reuse, recycle—only recycle has a realistic chance of succeeding in affluent American society. Three cheers for garage sales.

## Superfund

Thousands of toxic-waste dumps and landfills dot the landscape. Many of them contain hazardous materials in barrels that are rusted and leaking into local water supplies. About 70 million people, 25 percent of the American population, live near a toxic-waste site. Almost 1,200 public schools in five states, attended by more than 600,000 children, are located within 2,600 feet of a hazardous-waste site.[10] Epidemiological studies have correlated living near a hazardous-waste site with a significantly increased risk of several types of cancers and birth defects. The federal government estimates that there are 450,000 problem waste sites, not counting more than 7,000 toxic-waste sites at military facilities the government has protected from examination by the EPA.

In response to public pressure, the U.S. Congress in 1980 passed what has become known as the Superfund law. This legislation created a fund to pay for the identification and cleanup of contaminated sites. The money was mandated to come from the federal and state governments and from taxes on the chemical and petrochemical industries, who do most of the nonmilitary polluting. However, these industries vehemently opposed funding Superfund and in 1995 succeeded in getting Congress to repeal these taxes. Pre-1995 contributions by industry are now exhausted and taxpayers are funding 100 percent of Superfund's budget.[11] The government's reserve for cleaning up Superfund sites has been steadily depleted, declining from $3,789 million in 1996 to only $159 million in 2003.

There are 35,000 sites in the Superfund inventory, the most critical being assigned to the National Priorities List, which qualifies them for immediate attention. At the start of fiscal year 2003, there were 1,233 sites on the NPL with the greatest concentration in the industrial Northeast (figure 3.7). Remediation of NPL sites is slow, a result due largely to our legal system. As you can imagine, few companies admit liability and many sue everyone involved with the contaminated site. Studies indicate that about half of Superfund's money is spent on legal fees rather than cleanup efforts. A case often drags on through our clogged court system for many years. From the

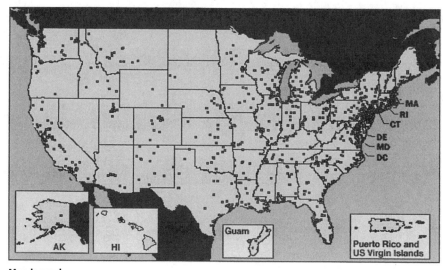

**Map legend**
☐ Abandoned hazardous waste sites (Superfund sites)

Figure 3.7
Superfund sites on the National Priorities List in November 2001. Since 1980, the
EPA has identified nearly 1,300 of these sites. The total number of sites in the
Superfund inventory exceeds 35,000 (Environmental Protection Agency).

point of view of a company's balance sheet, it is worth spending millions on
lawyers to delay spending hundreds of millions on cleanup costs.

The amount of money spent by federal, state, and local authorities on
pollution control and remediation is certain to grow during the twenty-
first century. The annual Superfund budget ranged from $1.3 billion to
$1.7 billion during the 1990s but was only $1.2 billion in 2002. As a classic song from the 1940s says, "You had your way and now you must pay.
I'll bet that you're sorry now."

Incineration
Incineration is potentially a better method of garbage disposal than landfills. It can generate energy while reducing the amount of waste by up to
90 percent in volume and 75 percent in weight. At present, 14–15 percent
of our solid waste is dealt with in this way, in about 400 incinerators designed for either municipal nonhazardous waste or for hazardous waste.
Those designed for hazardous waste require special devices to remove

toxic chemicals from gases produced before they exit the smokestacks. Unfortunately, solid-waste incinerators are, on average, substantially more expensive to build and operate than landfills, and sometimes are more expensive than recycling collection and processing. Despite the financial burden, more than 170 municipalities in the United States have decided to bear the comparatively high costs imposed by operating incinerators that have already been built.

Why have they done this? One answer is that the incinerator plants are already built and construction costs are a significant part of an incineration operation. Another explanation is that, despite environmental dogma, a wide-ranging study published in 1997 found that incineration may be environmentally friendlier than recycling. One reason for this surprising finding is the value of energy generated by incineration. Most of the 170 incineration plants in operation are designed to generate electricity by allowing the burning trash to heat water in a boiler to produce steam that drives a turbine to generate electricity. Another is that recycling uses large amounts of energy and creates pollution, especially when the solid waste is transported to the neighborhood recycling bin and from there to recycling mills. Much gasoline is burned for each ton of waste carried during these repeated trips. Then, the recycling mill itself uses large amounts of energy during processing of the waste stream.

There is a constant and acrimonious debate about the relative "greenness" of incineration and recycling. At present, the recycling enthusiasts are in the lead. They point out that many older incineration plants are major sources of atmospheric toxic materials such as mercury, lead, cadmium, and dioxins. Municipal incinerators produce more that half the dioxins emitted in the United States. If the poisonous materials are captured before leaving the smokestacks they end up in the ash, which must then be disposed of by encasing them in an inert solid material of some sort.

## Conclusion

Modern American civilization produces a very large and growing mountain of municipal solid waste. There is no sign that our annual production will decrease in the forseeable future. There are three ways to handle this waste, each of which has its pros and cons. Landfills can be designed to be safe, but most existing landfills are not, and even if they are they do not last forever. They fill up. New locations near the cities where most of the rubbish comes from are getting harder to find. Disposal costs are sky-high and growing for many cities, and the time may soon come when it will be intolerable for a city's budget.

Recycling and incineration are likely replacements for landfills but they also have drawbacks. Recycling saves resources but may be more polluting than incineration when all factors are considered. Recycling may make you feel good, but it is uncertain whether it is better than incineration. And recycling commonly is more expensive than operating a landfill. But this will change as the costs associated with landfills continue to rise. As "they" say, there will always be more people but there will never be more land.

One important advantage of recycling is that it avoids depleting dwindling natural resources. There is no doubt that many metal ores are becoming harder to find and that mineral extraction produces enormous amounts of environmentally unfriendly debris. In fact, the waste produced in mining and processing the metal ore produces 38 percent of the solid waste we generate in the United States. Water draining through these waste piles leaches metals such as arsenic, cadmium, and lead from the debris, polluting both surface and subsurface water supplies. EPA's Superfund list includes 77 mining-related sites on its National Priorities List.

There is disagreement about what cash value to put on resource depletion. And what is the environmental cost of each volume of carbon dioxide and particulates emitted by a moving vehicle taking recyclable materials to a bin or recycling plant? How often must a ceramic cup be used before the energy and materials used in its manufacture are less damaging to the environment than a plastic cup? What about the environmental damage done by using scarce water and detergents to repeatedly wash the ceramic cup, a process not needed for the throwaway styrofoam cups used in fast-food restaurants? Plastic bumpers on cars are less recyclable than steel but they are substantially lighter and improve gas mileage, reducing energy consumption and pollution. The controversies are endless and perhaps impossible to resolve to everyone's satisfaction.

Well-built incinerators reduce the volume of municipal solid waste by 90 percent but are expensive to build and operate. And the ash residue from an incinerator is rich in pollutant chemicals and must be treated before disposal. Our ability to recycle is limited by the way products are manufactured. An aluminum beverage can is pure aluminum and easy to reclaim and reuse in another product, but a metallic object that contains more than one metal creates large processing costs and is unlikely to be recycled. Incineration does not have these limitations. Despite its high cost the proportion of our waste that is incinerated will probably grow.

# 4

## Soil, Crops, and Food

Whoever could make two ears of corn to grow upon a spot of ground where only one grew before, would deserve better of mankind than the whole race of politicians put together.
—Jonathan Swift, *Gulliver's Travels*

Food ranks with clean water as essential for human existence, and for most Americans the concept of food scarcity is hard to imagine. Most of us are burdened by surpluses rather than shortages and our health problems have more to do with being overweight (61 percent, of whom 20 percent are obese) than with being hungry. Specialty manufacturer Goliath Casket Company now produces a triple-wide coffin, 44 inches across rather than the 24-inch standard, and their business is expanding (ouch!) by 20 percent annually.

We are the world's major breadbasket. We produce 25 percent of the world's food, and our waistlines show it. The National Center for Health Statistics says 15 percent of children between ages 6 and 18 were obese in 2000, compared with 6 percent in 1980, and experts believe the numbers are still increasing. Surgery to shrink the stomachs of obese Americans (bariatric surgery) increased 40 percent between 2001 and 2002 to 80,000, and the number is expected to climb to 120,000 in 2003.[1]

Compared to costs in the rest of the world, our food is almost free. Americans spend only 12 percent of their disposable income on food, a percentage that would be even less if we ate more meals at home rather than in expensive restaurants. Can America's agricultural abundance be maintained indefinitely? There are signs that even breadbasket America may be facing its productivity limit. The major reason for this looming catastrophe is our neglect of the soil in which our crops grow.

## The Nature of Soil

This nation and civilization is founded upon nine inches of topsoil. And when that is gone there will no longer be any nation or any civilization.
—Hugh Bennett

What is soil and where does it come from? Soil is born as rocks decay under the onslaught of rain and sun. The decay process produces the sediment we call soil, a mixture of clay and other minerals from the rotted rock. In addition to these grains, soils also generally contain a small percentage of black organic matter formed by organisms in the soil and the dead plants they feed on. One heaping tablespoon of soil may contain up to 9 billion microorganisms, 50 percent more than the human population on earth. So soil is a mixture of clay, other minerals, and organic matter. The organic matter is the stuff that makes the upper part of the soil dark and rich and good for growing vegetables, flowers, lawns, and crops. This black stuff also is responsible for holding most of the water in the topsoil. In an ideal agricultural soil the amount of organic matter is about 5 percent, composed of both living and dead plants and animals (bacteria, fungi, worms, and other small creatures).

Soils have a layered structure (figure 4.1). The layers or horizons are termed O, A, B, and C, from top to bottom. The A-horizon is also known as topsoil and this is the horizon in which our crops are rooted and live. It is the horizon we plow and fertilize and is usually only a few inches thick. It is hard to believe that so few inches of dirt nourish us as well as a large part of the world. Imagine 6 billion people dependent for their existence on an accumulation of a few inches of dirt. Clearly, such a thin earthskin is fragile and needs to be protected at all cost. How have we been doing at this job? By any reasonable standard, not well. As an academic, I would award America's effort no more than a C-, even with grade inflation. What are we doing wrong and how can we do better?

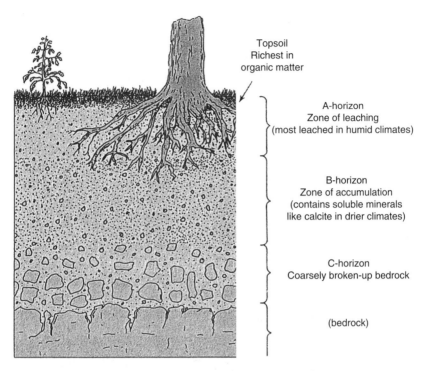

Topsoil
Richest in
organic matter

A-horizon
Zone of leaching
(most leached in humid climates)

B-horizon
Zone of accumulation
(contains soluble minerals
like calcite in drier climates)

C-horizon
Coarsely broken-up bedrock

(bedrock)

**Figure 4.1**
A generalized soil profile. Individual horizons vary in thickness and some may be locally absent. The A-horizon is dominated by mineral grains in various stages of decay, releasing plant nutrients in the process. Water and air are present between the mineral grains. The B-horizon is enriched in either red hematite (iron oxide) in humid climates or calcite (calcium carbonate) in dry climates.

## Farmland: Where Is It and Why?

A place for everything, and everything in its place.
—Samuel Smiles, *Thrift*

About 29 percent of America's land area is devoted to agriculture.[2] Where is America's farmland located and why there? There are three major requirements for a prime agricultural region: an extensive area of flat ground, a moderate to warm climate, and an adequate water supply. Where in the United States are such areas located? The regions that come to mind for extensive flatness and moderate to warm climate are the vast midcontinent of our country between the Appalachians and the Rocky Mountains, Florida, and the inland valley of central California (the San Joaquin Valley). So at first glance America is indeed blessed with agricultural land. The flat midcontinent alone covers about half the 48 contiguous states.

But there are problems with using some of this flat land, the major one being not enough precipitation. The best farmland needs 30–40 inches of rainfall each year to reach its maximum potential, and both the midcontinent and the San Joaquin Valley fall considerably short of this goal. The annual precipitation averages 15–30 inches in the midcontinent (figure 1.1); in the San Joaquin Valley, it averages only 8–12 inches. Our largest flat area is thus marginal in terms of precipitation, and central California would seem an unlikely place to locate agricultural abundance. After all, 8–12 inches of rain is typical of southern Arizona and not many people think of that area as an agricultural breadbasket. But extensive crops are grown in both the midcontinent and southern Arizona because of underground aquifers, aquifers that are fast becoming depleted, as we saw in chapter 1. Insufficient aquifer water is rapidly becoming a major problem for America's agricultural community.

## Urban Growth and Farmland Loss

The enormous superstructure of American society rests on the tiny point where fewer and fewer farmers with larger and larger machines mine as much food as they can from fields that are less and less what they once were.
—Jane Smiley

They paved paradise and put up a parking lot.
—Joni Mitchell, "Big Yellow Taxi"

Impending water shortages are only one of the problems facing America's farmland. Another is urban sprawl, which is moving ever faster with each passing year. From 1982 to 1992, an average of 2.5 million acres of farmland was lost each year to urban development. From 1992 to 1997, the rate was 3.2 million acres per year. The number of people living in our cities rose 80 percent between 1950 and 1990, while the land covered by those areas expanded 305 percent. These lost farm acres stopped sprouting grain and instead sprouted housing developments, shopping malls, and other construction. In California, where half of the nation's fruits and vegetables are grown, 16 percent of the best soils now underlie urban areas, as do 9 percent of the next-best soils.[3] There is 20 percent less farmland in the United States today than in the 1960s because of commercial development. Commercial developments not only absorb farmland but also drive up land prices. This provides a large financial incentive for financially stressed farmers living nearby to "take the money and run," further decreasing the amount of farmland. The development process feeds on itself to eliminate farmland.

The decrease in farmland caused by people moving to cities for increased opportunities is made even more serious by the increasing average age of America's farmers, up from 51 in 1970 to 57 in 1998. The average age of all U.S. workers is 38. Thirty-five percent of farmers are now 65 or older, compared to 13 percent of all Americans.[4] Nationwide, there are 300,000 fewer farmers than there were 20 years ago. The financial hardships of running a farm (predicted to get worse with global warming), a decline in the size of the average farm family, and increasing opportunities in other fields mean that many farmers will not be replaced when they retire. The farmland will then probably be purchased by commercial developers for more suburban housing. Developers prefer land with deep, well-drained, and nearly level soils, areas best suited for agricultural production. Thus, the retirement of farmers not only leaves fewer people producing food, but also less farmland.

Urban sprawl has another bad effect on farming, but a more subtle one than the obvious loss of farmland. When people concentrate in cities air pollution from cars and factories increases significantly, which causes a significant decrease in crop yields (table 4.1). The smoggy haze decreases the amount of sunlight available for photosynthesis, and the chemical soup in the haze reduces a plant's growth and fruit yield and increases its susceptibility to disease and insect attack.[5]

As a result of both the decreasing amount of farmland and increased mechanization, only 2.6 percent of Americans work on farms nowadays; the number of farms has decreased by 50 percent but average size has

Table 4.1
Estimated crop losses in southern California, determined by growing crops in filtered, unpolluted air and in smoggy air

| Crop | Percent |
| --- | --- |
| Alfalfa | 8.6 |
| Beans | 10.3 |
| Cotton | 18.6 |
| Grapes | 28.1 |
| Oranges | 26.2 |
| Potatoes | 14.9 |
| Rice | 5.3 |
| Tomatoes (processing) | 3.8 |

*Source:* California Air Resources Board, 1991.

doubled in the last 50 years.[6] The 8 percent of farms with sales above $250,000 occupy about 27 percent of farmland but generate 68 percent of total farm output.[7] They also get the lion's share of the $19 billion in annual farm subsidies from Washington. There are still a large number of "mom and pop" farms but most are not profitable and their numbers are decreasing. They are becoming a nostalgic and romanticized vestige of America's past.

## Soil Erosion

Man has always found it difficult to appreciate the delicacy of quality and texture of soil, and to realize that it is not inexhaustible, and that some processes are virtually irreversible.
—Anthony Huxley

The "muddy Mississippi" is a stereotype in the American consciousness but few of us think about where the mud comes from. It takes but a moment to realize that it must be dirt sloughed off from the land bordering the river—mostly farmland. And a lot of it is the precious topsoil in which we base our agricultural abundance. Some of American agricultural land may be losing topsoil faster than nature can reform it, but it is not clear whether the average rate of topsoil loss is too great to be sustainable. The average rate of soil erosion in American agricultural areas is hotly debated among the experts, but all agree it is a problem that needs to be addressed. Available data are not good enough to determine whether the loss of topsoil is increasing, decreasing, or remaining constant. Similarly, the relationship between the rate of loss of topsoil and its formation by natural processes is also uncertain. In the climate of the American midcontinent it takes many decades or perhaps hundreds of years to form 1 inch of agricultural soil.

## Why Does It Happen?

The leveled lances of the rain at earth's half-shielded breast take glittering aim.
—Paul Hamilton Hayne, *A Storm in the Distance*

The major culprit is raindrops smashing into loose dirt. A large raindrop falling in still air can reach a speed of 20 miles per hour, and when the rain is wind-blown the speed is even greater. Every drop of water is a hammer dislodging and moving soil particles down even the gentlest slope. When we take into account that millions of raindrops fall in a typical rainstorm the potential for soil movement downslope toward the nearest stream is easy to see.

About one-third of soil movement results from wind blowing across the land surface. In some areas of America's breadbasket the wind is a more potent soil-moving force than water, sometimes four to five times more potent. The midcontinent is America's windiest flatland. This was disastrously brought home to farmers during the infamous dust-bowl years of the 1930s. The combination of soil loosening by the plow and lack of rainfall during an extended drought left the soil surface exceptionally vulnerable to movement by the wind, a circumstance that continued for several years. As we noted in chapter 1, annual precipitation in our largest farming area is far below the 30–40 inches that is optimum for agriculture, averaging perhaps 20 inches over the region. In about one year in five the annual precipitation drops to 15 inches, virtually a recipe for crop failure unless the rainfall is supplemented by irrigation water from underground aquifers. There was not nearly the amount of irrigation in the 1930s that there is today, so the farmland was drier. A dust-bowl phenomenon is much less likely today than it was 70 years ago.

## How Can Soil Erosion Be Stopped?

The worst farming practices use up eight inches of topsoil (which took 7,000 years to create) in 36 years. The best practices would make that eight inches last 2,224 years.
—Iowa Department of Soil Conservation

When new crops are ready to be planted using conventional agriculture, the remains of last year's crop are plowed into the soil, exposing a bare soil surface. Then the bare soil surface is broken up to make a smooth bed for seeds. Finally, after the seeds are planted the soil between the rows is stirred to rip out weeds. Thus, the topsoil is repeatedly worked over, granulated into tiny fragments, and any existing soil-holding roots from the previous crop are destroyed. The soil is left unprotected until the new seeds germinate and develop roots, which takes several months. The soil is not protected from being carried away by rainwater during the spring rains. Standard cultivation practices increase soil erosion by a staggering 10 to100-fold.

### Conservation Farming

A better planting method now used by most farmers is called conservation farming. Two principles are central to this farming method: (1) reduce the amount of soil disturbance—that is, plowing; (2) do not leave the soil bare and exposed to wind and rain. The amount of plowing and disturbance of the soil is kept to a minimum. There may be no plowing of the soil at all. Seeds are planted directly over the residue of last year's crop, regardless of whether the new crop is the same or different from the previous one. This way nearly all the soil remains protected during spring rains. In addition, moisture is retained in the soil by living plant roots, holding the soil together and reducing runoff of water and topsoil from the field. Soil loss is reduced by more than 90 percent. Another benefit of conservation farming is that valuable soil organisms such as earthworms are not turned into wormburger meat by the plow. Experience has shown that no-till farming results in higher crop yields, with lowered use of water, pesticides (weed seeds are not brought to the surface), and tractor fuel.

**Contour Plowing**

When the ground has a significant slope a technique called contour plowing should be used. Ridges and furrows are constructed following the con-
tours of the ground slope so that the downhill flow of water is reduced.
Even a small decrease in flow velocity produces a large decrease in erosion.
Contour plowing is effective on slopes of less than 8°. On steeper slopes
terracing is effective, but America's extensive open spaces for agriculture
make this farming method unnecessary.

**The Need for Trees and Forests**

Trees serve two functions for farmers. In windy country such as the American Midwest, lines of trees act as a windbreak to reduce wind erosion.
Perhaps more importantly, trees and their root systems on slopes hold water and are essential to prevent flooding of nearby agricultural fields during heavy rains. Forested areas also prevent slope erosion and smothering
of cropland margins by sediment moving rapidly downslope.

**Soil Pollution**

All this is not to say there is no insect problem and no need of control. I
am saying, rather, that . . . the methods employed must be such that they
do not destroy us along with the insects.
—Rachel Carson, *Silent Spring*

A caterpillar on your cauliflower is a sign that both are safe to eat.
—Peter Thomson

Fertilizers of some sort are always applied to crop-growing land to increase plant health and productivity. Perhaps most common are the commercial NPK (nitrogen, phosphorous, potassium) products, the elements
needed in greatest quantity by most plants. These are universally agreed
to be good for crops. However, an analysis of 29 commercial fertilizers by
the U.S. Public Interest Research Group revealed that each of them contained 22 toxic heavy metals. In 20 of the products, levels exceeded the
limits set for waste sent to public landfills.[8]

More contentious is the use of sewage sludge, about whose health risks there is "persistent uncertainty," according to the National Academy of Sciences. Before passage of the Clean Water Act in 1970, most raw sewage was dumped into oceans or rivers. Now almost everything flushed down toilets and poured into drains by industrial plants, hospitals, gas stations, and householders eventually finds its way to a wastewater-treatment plant. The water, with toxins removed, is treated and returned to waterways. The rest, sewage sludge, may be treated to a greater (more expensive) or lesser (less expensive) degree. The more expensive treatment produces sludge with the least health risk but is not as rich in plant nutrients as the less expensive treatment. But the less expensive treatment leaves more pathogens in the sludge. Farmers get sludge free, and in some cases may even be paid to take it.

About 5.6 million tons of sewage sludge, commonly called biosolids in environmental literature, is generated each year in the United States. More than half of it is recycled into our soil, and the rest is buried in landfills or burned.[9] According to the National Sludge Alliance, sewage sludge contains 60,000 toxic chemicals, a mix of residential, industrial, and commercial discharge that includes hospital wastes, street runoff, heavy metals, PCBs, dioxin, solvents, asbestos, and radioactive wastes.[10] The EPA published purity standards for sludge in 1993 that many scientists now consider inadequate, but there have been no scientific investigations or documentation of health impacts by the EPA. Numerous lawsuits have been filed by people who believe their health has been compromised by toxins in sludge.

Nearly all of America's 2.2 million farms today use herbicides and insecticides to control unwanted insects and plants ("weeds"). Most farmers consider these products essential to maintain productivity and apply 1 million tons of them to their fields each year. But less than 1 percent of applied pesticide reaches the target pests.[11] The FDA has determined that at least 53 carcinogenic pesticides are presently applied in massive amounts to our major food crops. And in 1998 they found that 35 percent of U.S. food samples contained pesticide residues. Also, because the chemicals applied to the crops dissolve in water they cannot be contained within the boundaries of the farm. They drain outward into adjacent waterways as well as downward into groundwaters. Pesticides have become 10 to 100 times more toxic than they were 30 years ago, and their use results in between 3.5 million and 5 million acute poisonings each year. Farms produce 70 percent of the stream pollution in the United States.

There are no health data for many of these chemicals, although the EPA in cooperation with the chemical industry is now testing large numbers of previously untested chemicals for general toxicity and for evidence of endocrine disruption. However, many chemicals have not been tested on humans to determine possible long-term effects and, because many new chemicals are devised each year, such testing can never be completed. Thorough testing takes a great deal of time, but it takes little time for chemists to concoct new poisons.

Perhaps even more frightening is the realization that chemicals ingested in combination can be many times more harmful than the chemicals are individually.[12] Yet chemicals are still only tested by the EPA for their carcinogenic or mutagenic potential in isolation from each other. But we face an unsolvable testing problem. To test just the commonest 1,000 toxic chemicals in combinations of 3 at a standardized dosage would require at least 166 million different experiments. And what about different dosages?

Even ingredients listed on pesticide labels as "inert," commonly more than 90 percent of a pesticide product, may not be safe for humans. According to a survey by the Northwest Coalition for Alternatives to Pesticides, about a quarter of inert substances are classified as hazardous under the Clean Air Act, the Safe Drinking Water Act, and other federal statutes.[13]

In May 2001, the United States signed the Convention on Persistent Organic Pollutants (POPs), committing us to reduce and/or eliminate the production, use, and/or release of the 12 POPs of greatest concern to the global community. The treaty also established a mechanism by which additional chemicals can be added to the list in the future as new dangers are recognized. Most herbicides are not harmful to humans but can be toxic to fish and other wildlife. Many types of insecticides, however, harm not only birds and fish but humans as well. Studies show that farmers who work with pesticides get Parkinson's disease and several kinds of cancer more often than the general public does. Each year thousands of people, most of them not farmers, are admitted to hospitals in the United States for pesticide poisoning. Because of increasing fear of pesticides, the use of these chemicals has declined somewhat over the past 20 years but they are still considered staples in most American farming. Because of pesticide pollution, loss of soil fertility caused by conventional plowing practices, soil erosion, and the increased incidence of pesticide-resistant crop diseases, it is unclear whether farming as now practiced in the United States can be sustained indefinitely.

It is worth noting that insect pests seem able to develop resistance to new pesticides as fast as the pesticides are developed by agricultural chemists. In 1950 there were about 10 species of insects and mites resistant to pesticides; today the number approaches 600. The number of weeds and plant diseases with this resistance has increased from near 0 in 1950 to more than 400 today.[14] Nevertheless we hear from Monsanto and the other pesticide manufacturers that newer pesticides are more potent than older ones.

Even though the amount of insecticides used in the United States increased tenfold between 1945 and 2000,[15] 37 percent of preharvest crops were lost to insects in the 1990s, compared to 30 percent in the early 1940s. The percentage of the crop harvest destroyed by pests in medieval Europe was 30 percent.[16] Today the world average loss is 35–42 percent, despite the increasing use of pesticides in recent decades. Are pesticides making our pest problem worse? If they are, not everyone will be saddened. In Japan, a memorial service has been held by the Society of Agricultural Chemicals Industry annually for the past 40 years to honor the memory of insect "victims" of agrochemicals.[17]

### Integrated Pest Management

A minority of farmers, wary of pesticides, have sharply reduced their use, preferring a system of pest control called *integrated pest management*. Crops are rotated yearly from field to field to disrupt pest infestations. Conservation farming is practiced to build fertility and reduce the need for expensive fertilizers, most of which contain toxic levels of heavy metals. Predatory organisms are released into the fields to control harmful insects. Pesticide use is not entirely abandoned but used only if pests reach a threshold level. Studies of integrated pest management in Britain showed increased harvests, and farmers' profits increased by 20 percent.

## Organic Farming

What is a weed? A plant whose virtues have not yet been discovered.
—Ralph Waldo Emerson

Some farmers have eliminated all use of synthetic pesticides and fertilizers, a process known as *organic farming*. While the total number of farmers has been falling, the number of organic farmers, now at 12,000, has been rising steadily in response to increasing public concern about pesticides in food. Younger farmers are more likely to farm organically. The average age of organic farmers is 10 years less than that of other farmers. There was a 15–20 percent increase in organic acreage each year during the 1990s. About one-third of Americans buy some organic products. Between 1980 and 2001, organic farming revenues climbed from $78 million annually to $9–9.5 billion of a $500 billion grocery market.[18]

Only 0.3 percent of crop acreage in the United States is farmed organically.[19] On these farms, the use of synthetic fertilizers has been abandoned in favor of animal manure and crop residues. As in integrated pest management, crops are rotated and predatory insects such as ladybugs (are any of them males?) are introduced to control harmful pests. Soap may be sprayed on plants to protect them rather than insecticides. In the European Union a certain species of naturally occurring bacteria is sprayed onto wheat, barley, and oat seeds to combat fungal diseases. It is nontoxic and has proven 98–100 percent effective.

Organic fields that have been under a sustainable, fertility-building agricultural regime for many years—or whose fertility has never been "drawn down" by chemical applications and repeated monocultures—can outperform industrially farmed ones.

They use much less commercial fertilizer, half the energy per unit yield, and increase the number of nutrient-cycling microbes, worms, and helpful fungi. In the U.S. Midwest, farmers who produce grain and soybeans organically are finding that their net profits equal or surpass those from conventional production, even when they do not charge the premium prices that organic crops generally command.

However, it takes 5–10 years of sustained organic farming to eliminate the pollution caused by decades of pesticide applications and, until these poisons are washed from the soil, crop yields from organic farming are lower than those obtained using chemical sprays. An added benefit of eliminating the use of pesticides is that organic systems are more resilient in maintaining productivity in drought years that lead to disastrous failure in conventional agriculture.

In response to growing public interest in organic food, the U.S. Department of Agriculture earmarked $5 million in the 2001 budget for research on organic food production. In October 2002, the USDA implemented a voluntary labeling program for organic food sold in stores. The label "100 percent organic" certifies that the food was grown without pesticides, hormones, antibiotics, irradiation, or genetic modification. "Organic" means at least 95 percent and "made with organic ingredients" means at least 70 percent. Products containing less than 70 percent organic ingredients can identify organic ingredients on their ingredient list.

Food grown without pesticides contains substantially higher concentrations of 21 nutrients[20] as well as increased amounts of antioxidants and other health-promoting compounds compared to crops produced with pesticides.[21] Heavy use of pesticides and chemical fertilizers seems to disrupt the ability of crops to synthesize certain chemicals that are associated with reduced risk of cancer, stroke, heart disease, and other illnesses. This discovery is another piece of evidence demonstrating that messing with the environment in which we evolved is a bad idea and should be avoided whenever possible, which is most of the time.

About a third of medium-priced restaurants now offer items they term organic on their menus. The nation's largest organic grocery chain has a growth rate of 30 percent, about double that of the supermarket industry in general. One out of every four Americans buys organic products. Gerber, Heinz, General Mills, Nestlé, and Unilever have now entered the organic produce market with baby food, flour, and other products. Almost half of American consumers say they are interested in purchasing organically grown products, now available at 73 percent of grocery stores. A newly founded city in Iowa has banned the sale of nonorganic food within the city limits.

Organic farming is promoted by governments outside the United States. The Israeli government gives subsidies to organic farmers. At present only 1–2 percent of their population buys organic food. Sales of organic food items in Britain rose 50 percent in 1999 compared to 1998, and rose another 50 percent in 2000. In Western Europe the amount of land under organic production is now more than 3 percent of total EU agricultural area. In several European nations 5–10 percent of total agricultural area is organic. The current record holder is Austria with 10.4 percent.[22] In 2003, ministers from the European Union countries published an action plan to promote organic farming. In 2003, the Ethical Exchange Management Company, based in London, opened the first international commodities exchange devoted entirely to organic products. Cuba has adopted organic farming as the official government strategy for all new agriculture in Cuba, after its highly successful introduction in 1990.

### The Cost of Purity

An important question for American consumers is price. Organic food costs more in the stores, 57 percent more on average than nonorganic food. Why is this? There are several reasons, not necessarily related to the cost of growing the crops. Conventional farms are generally larger than organic farms and are heavily subsidized by the federal government—that is, nonorganic produce does not carry its full cost. American taxes subsidize it. However, this inequitable financial burden on organic farming may be changing. The U.S. Department of Agriculture budget has included an organic crop insurance program since 2001.

Another reason organic foods cost more in stores is that the organic produce infrastructure has been too small to benefit from economies of scale. However, the organic products food sector is now growing at a rate of 25 percent a year, which has resulted in the emergence of large organic food supermarket chains such as Whole Foods, Trader Joe's, and Wild Oats. Whole Foods has 143 stores in the United States and Canada and offers more than 1,200 items. Wild Oats competes with 102 stores.[23] Most or all of the difference in cost between conventional and organic food products would disappear if organic farming were a larger-scale operation and were able to reap the financial benefits that accrue to large farms under existing federal agricultural policies.

## Genetically Modified Foods

The control of nature is a phrase conceived in arrogance, born of the Neanderthal age of biology and philosophy, when it was supposed that nature exists for the convenience of man.
—Rachel Carson, *Silent Spring*

A revolution in commercial agriculture began in 1994 with the introduction of the Flavr-Savr tomato, engineered for delayed ripening. Two years later genetic engineering took off and now is spreading like wildfire. Large areas of genetically modified (GM) or transgenic crops seem to be taking over America's farmland. Extensive fields of GM soybeans, corn, cotton, and canola are now being grown in the United States. In 2002, 80 percent of America's soybean acreage was planted with GM herbicide-resistant seeds. This compares to 75 percent in 2001, 54 percent in 2000, 47 percent in 1999, 12 percent in 1997, and only 2 percent in 1996. In fields planted with the genetically modified soybeans weeds were more easily controlled, less plowing was needed, and soil erosion was minimized. GM corn is now 38 percent of all corn being grown; it produces its own insecticide. GM cotton in 2002 was 70 percent of cotton grown, up from 61 percent in 2000 and 48 percent in 1999; it is engineered to tolerate herbicides. Transgenic potatoes, tomatoes, melons, beets, and other crops have been approved for production by the U.S. Department of Agriculture. The operative phrase seems to be "full speed ahead."

Rice, the world's number one food staple, has been turned into yellow-colored "golden rice" in greenhouse experiments (gene from a daffodil). This GM rice yields a crop 35 percent larger than normal rice, has built-in resistance to pests, and contains vitamin A and iron, substances lacking in normal rice. Yellow rice is touted as a way to end the loss of vision that results from lack of vitamin A as well as a way to end most iron-deficiency anemia among rice-loving Asians. However, the average person would have to eat 20 pounds of cooked rice every day to get the necessary vitamins.[24] Yellow rice will soon be available for planting; seeds will be distributed free of charge to farmers in poor countries such as Vietnam, Thailand, and Bangladesh. Proponents of GM crops see the rice as the solution to feeding the coming world population of 8 or 10 billion people without destroying additional land, forests, and water.

The latest twist in genetically modified crops is called "biopharming," the production of "edible vaccines." In biopharming, crops and animals (e. g., milk from cows) are turned into factories to make drugs. Building vaccines into cornflakes or other foods could be especially helpful for developing countries, where syringes, refrigeration, and trained medical personnel often are scarce. One company has added a virus gene to corn to create corn that protects the consumer from hepatitis B.

GM crops are produced by transferring a few selected genes from one organism to another to produce a specific useful effect. Hundreds of variously modified foods are now in the supermarket pipeline, with genes borrowed from every form of life—bacterial, viral, insect, even animal. Bacterial genes can be put in corn; fish genes can be inserted into tomatoes. A gigantic game of "mix and match" is in progress. As of 2002, more than 50 different "designer" crops have received federal approval, and about 100 more are undergoing field trials.

The U.S. produces more than 70 percent of the world's genetically engineered crops. Two-thirds of the products on supermarket shelves contain genetically modified ingredients such as soy or corn.[25] We have all been eating genetically modified food products despite the fact that only 25 percent of American consumers believe GM plants are safe and more than half of Americans say they do not want to eat GM food. The U.S. Congress has yet to pass a law requiring labeling of foods that contain genetically modified ingredients, despite a recent poll showing that nearly 92 percent of the public wants such labeling.[26]

### Objections and Fears Concerning GM Crops

Tampering with nature always involves risks. A growing number of studies have suggested that GM crops could lead to rapid evolution of pesticide-resistant insects, creation of new plant diseases, weeds that can no longer be controlled, and harm to insects that benefit humankind.

Controversial laboratory investigations suggest that physiological changes in humans may result from seemingly innocuous gene manipulations in the food we eat. We may be creating Frankenfoods. A national poll by the Pew Research Center in 2002 revealed that 55 percent of Americans (47 percent of the men, 62 percent of the women) believed that genetic engineering will upset the balance of nature and damage the environment; 37 percent believed this was not likely. Democrats are more concerned than Republicans, 58 percent to 51 percent.

What is particularly worrisome is that because biological systems reproduce, genetic "pollution" cannot be cleaned up like a chemical spill or recalled like a defective automobile. Once the gene, or genie, is out of the bottle, so to speak, it cannot be put back in. And the genie is out.

Europeans have been particularly anxious about eating GM foods; 94 percent want them labeled (they were in 2003) and 70 percent object to eating them.[27] They were outraged by the announcement in May 2000 that genetically modified seed was accidentally planted on several farms in Europe. A company in the United States says the inadvertent planting of GM seeds is probably commonplace. The company, Genetic ID, screens agricultural products for GM material. They found that 12 of 20 random samples of conventional seed taken from American distributors contained some GM seed. Surveys in 2000 of grain handlers found that 80–90 percent intended to purchase biotech crops, but only 10 to 25 percent will segregate crops.

In 1999, Canadian farmers reported weeds that had acquired herbicide tolerance from neighboring transgenic crops just 2 years after the crops had been planted. Stray pollen and seed from GM oilseed rape, or canola, is now so widespread in Canada that it is difficult to grow conventional or organic strains without them being contaminated. In 2001 pollen from genetically modified maize was found in remote mountainsides of Mexico, having been blown there from plots of GM maize 60 miles away. In Hawaii in 2003 genes from an experimental crop of bioengineered corn spread to other corn growing nearby.[28] Common sense suggests that it is not possible to build a wall high enough to keep genetically modified organisms out of the environment, because pollen often drifts for miles on the wind, potentially contaminating everything in its path. Farmers are being sued by Monsanto for inadvertently growing GM products without permission. The farmers are, reasonably enough, outraged that they are being held responsible for blowing wind.

In rebuttal, supporters of GM crops say the critics ask the impossible: technology without risks and a promise of unconditional safety. They point out that there can never be a guarantee that anything is harmless. But unless there is evidence of harm, we shouldn't worry. Without risk taking there can be no experimentation, and therefore no progress. There is no such thing as total proof, no such thing as zero risk.

The supporters of transgenic crops say there is no difference between genetic modifications of plants by classical breeding methods to produce, for example, hybrid corn and using gene transfer to accomplish a similar result. If anything, they say, genetic engineering is more precise because it introduces just one or two genes into a plant. With conventional breeding, thousands of unknown genes are transferred in order to get the one with the desired trait. If altering genetic makeup is "unnatural," they say, humans have been performing unnatural acts for thousands of years. You don't find poodles or seedless grapes in nature. With the assortment of genes now known, crop designers can simply choose the traits they want and impart them in one step instead of by trial and error. Simply replace a gene believed to be neutral for humans with one that is beneficial.

The contamination problem may be particularly worrying in regard to pharmaceuticals produced from GM crops. Although some drugs you might ingest without realizing it would be digested before entering your bloodstream, others would not. Some oral drugs, such as plant-derived birth-control hormones, would not be digested and could cause havoc if they found their way into food. As one health expert has put it, "Just one mistake by a biotech company and we'll be eating other people's prescription drugs in our cornflakes."

The most extreme skeptics fear we may lose ourselves in the rush to play God. They point out that biotechnology deals with information coded chemically in living cells. Tools are being developed that enable us not only to decode this information that's encoded in DNA, but also to change it. This has never before been possible. We may be charging into a dark room that has no exit after we enter. Are GM crops another example of technological abilities outstripping wisdom?

As I write this late in 2003, there seems to be a growing resistance in some parts of the American business community to genetically modified foods. Several major companies—including Frito-Lay, McDonalds, Gerber, and IAMS pet foods—have said they will no longer use genetically modified ingredients in their products. Major producers of baby food, in response to consumer preference, have stated they will not use them either. Pets and babies are sensitive issues in the American marketplace. And some major grain handlers have announced they want genetically engineered corn and soybeans kept separate from unmodified varieties, although this is probably impossible to achieve with 100 percent accuracy. Genetically modified food is the most contentious issue in American agriculture today.

It is too late to debate whether genetically modified food is a gift from God or a spawn of the Devil. It is here to stay and will likely be more widely accepted as the years pass. After years of investigation, there is no convincing evidence that GM crops pose risks to human health or that they will lead to an ecological breakdown. However, it is a good idea to be vigilant about GM experimentation, so that excesses in its use do not occur. In 2001, lawmakers in 36 states introduced 130 pieces of legislation dealing with agricultural biotechnology. Twenty-two state legislatures passed bills dealing with GM crops.[29]

## Salt Accumulation In The Soil

And the whole land thereof is brimstone, and salt, and a burning, that it is not sown, nor beareth, nor any grass groweth therein.
—Deuteronomy 29:22

All natural waters contain dissolved salts obtained from rocks at the earth's surface. When the water evaporates, these salts precipitate and form mineral accumulations at and immediately below the soil surface, a process called salinization that affects 6 percent of America's farmland. It is a problem mostly in heavily irrigated areas such as the Western United States, where low annual precipitation necessitates extensive irrigation. In the San Joaquin Valley of California, for example, reports from more than a century ago tell of farmland abandoned because it became too salty for crops. Today, in California's Imperial Valley, more land is being taken out of production than is entering because of salinization. In Utah, a highway outside Salt Lake City offers a view of barren former croplands now crusted with poisonous white salts. Salinization is affecting more and more of America's cropland.

There is no easy or perfect solution for soil salinization. Flushing the soil with large amounts of low-salt water is very expensive and only slows the salinization process. And the flushed salty water simply migrates downstream to plague another farmer. Further, the flushing must be done slowly, for if the soil is flushed more rapidly than the salty water can drain away, the soil becomes waterlogged and unsuitable for planting.

Perhaps the best hope for taking the salt out of salinized soils is genetic modification of plants. A gene has recently been discovered that not only enables plants to withstand extreme salinity, but also helps them draw salt from the soil. Crops with a more active version of the gene could rehabilitate land lost to salinization.

## Cleaning Polluted Soil

What we need is more people who specialize in the impossible.
—Theodore Roethke

Cleaning organically polluted soil is expensive, takes much time, and is difficult under the best of circumstances. As we have already noted, it may take a decade to restore the vitality of insecticide-damaged soil. Other pollutants can last almost indefinitely—for example, spilled hydrocarbons (petroleum products such as gasoline). Perhaps the most promising method at present for cleaning polluted soil is the injection of contaminant-eating bacteria. Some exist naturally in soil and are quite specific in their tastes. Different organisms attack oil, raw sewage, treated sewage sludge, and other wastes. Within a year or two the amount of contaminant can be reduced to safe levels. Biologists are now creating genetically modified soil organisms that find particular contaminants particularly tasty. There is a glimmer of hope for restoring organically polluted soil.

## Agriculture and Climate Change

Climate is what you expect. Weather is what you get.
—Robert A. Heinlein

There is more carbon dioxide in the air now than at any time in the past few thousand years. Rainfall in the United States has increased significantly during the past hundred years. And the climate is getting warmer. What do these changes mean for American agriculture? The answer at present is uncertain. On the plus side, more carbon dioxide means more photosynthesis. Rainfall also stimulates plant growth. Warmth is likewise good for plants. And the effects of these climatic changes are already visible. Satellite data from 1980 to 2000 reveal that plant density has increased significantly. Data from northern latitudes between 42° latitude (where Boston and Chicago are) and 70° latitude (northern Canada) indicate that the growing season in Alaska has lengthened by 17 days since 1950. Spring is arriving earlier and summer is ending later. At lower latitudes (most of the midcontinent grain belt) the lengthening is only 12 days[30] because the temperature change has been less. But the trend seems clear—more plant growth.

Unfortunately, some of the negative factors that influence plant growth and crop production are also increasing. Weeds (plants you don't want) also benefit from increased carbon dioxide, rainfall, and temperature. So do insect populations and plant fungal diseases. As a result there may be an intensified cry for more herbicides and insecticides or more crops genetically modified to resist insects and fungi by farmers and perhaps a resulting decrease in the growth rate of organic farming. An added negative factor for crop production is an observed increase in extreme precipitation events and resulting flooding and crop damage. Exactly how the pluses and minuses for agriculture associated with climate change will balance out is not yet clear. Data are still too few for firm conclusions to be drawn.

## Conclusion

A few inches of dirt is all that separates us from mass starvation. We need to protect and preserve this layered surficial resource we call soil at all costs, but many or most farming practices concentrate on short-term profit rather than long-term survival. Unnecessary erosion, deliberate pollution by artificial chemicals, and salinization have been going on for many decades and now threaten the productivity of breadbasket America. Unless these trends are reversed we, as well as the many nations that depend on us for food, are in deep trouble.

The amount of land being farmed in the United States has been continually decreasing since 1950. Part of this decrease has resulted from increasing urbanization and suburban sprawl, phenomena that show no signs of stopping. As they say, money talks, and when a commercial developer offers big bucks for farm property, farmers are no different from the rest of us. They are likely to sell. Another factor in the decrease of farmland is the trend of young people leaving the farm for the lure of the big city. Farming as a "calling" is not as strong a factor as it once was for the children of farmers. The average age of farmers is increasing.

The effects of genetically modified crops on the future of farming are not clear. There are recognized benefits but also some reasons for concern. But the genie is out of the bottle. Experiments with transgenic crops are certain to continue, and we can only hope that the light of better crops to come is not the light beam of an oncoming train.

Most of the changes associated with our increasingly warm climate probably will be beneficial for American agriculture. But it will not be all sunshine and roses. Some plant pests and diseases are likely to become more bothersome and may spur a call for more herbicides and insecticides. Or perhaps for more crop varieties genetically modified to resist these invasions. American agriculture has entered an era of rapid and perhaps fundamental change. It cannot be stopped. We hope it can be controlled.

# 5
# Energy Supplies

How can the present market, which determines the price of electricity, coal, and so on, possibly represent the true cost of these fuels or the energy produced from them if there is no charge for the environmental damage they do?
—Stephen H. Schneider, *Global Warming: Are We Entering the Greenhouse Century?*

Everything contains energy. If you want to extract and use this energy to power an industrial civilization, you have to find sources that are abundant and from which the energy can be extracted commercially using available technology. Fortunately, on our planet there are a large number of choices, but each has both advantages and disadvantages. The bulk of modern industrial civilization is based on the fossil fuels oil, natural gas, and coal (figure 5.1), because they are abundant and easy to use. But they are terrible pollutants. Nuclear energy comes with fears about cancers and the accidental escape of dangerous radiation. Solar power is inexhaustible, but it has proven difficult to capture a large percentage of the sun's rays and convert them into electricity. Likewise, although the wind is strong enough in many areas to serve as a power supply, its speed is very variable from day to day and methods of storage are poorly developed.

The Department of Energy forecasts that world energy consumption will rise by 58 percent between 2003 and 2025,[1] despite continuing increases in energy-use efficiency, which more than doubled between 1920 and 2000.[2] America (and the world) have four choices. We will either (1) have to increase our use of the current major energy sources, oil, natural gas, and coal; (2) enlarge our use of currently minor sources (nuclear energy, hydropower, geothermal power, and biomass); (3) increase current trace sources (solar power, wind power) by at least an order of magnitude;

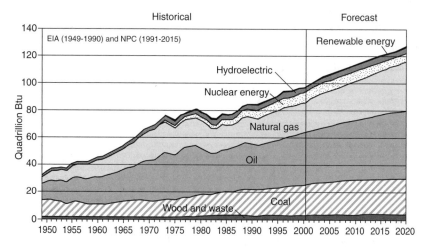

**Figure 5.1**
Trends in energy usage since 1950 and the projection until 2020 (Energy Information Administration).

or (4) develop entirely new sources of energy, such as fuel cells (hydrogen power), wave power, or tidal power. What should we do? What is the best mix of energy sources for America?

Part of the answer to this question is relative cost. Americans pay an average of 9.7 cents per kilowatt-hour for electricity,[3] the exact amount varying with geographic area and the mix of energy sources used by the local electric utility. Current costs for the sources of energy used to generate electricity are shown in table 5.1.

## Fossil Fuels

If America remains addicted to oil, Saddam Hussein will not be the last oil dictator, and this will not be the last oil war.
—Mike Clark, 1991

Fossil fuels are sources of energy formed millions of years ago from decomposed and chemically altered remains of dead animals and plants. Petroleum and natural gas were originally the organic remains of microscopic organisms that live at the ocean surface. Coal was formed from land plants. All three fossil fuels take millions of years to form after

Table 5.1
Costs of electricity (cents per kilowatt-hour), with and without external costs. External costs are estimated environmental and health costs for 15 European countries. Note the degree to which relative ranking changes when external costs are included.

| Electricity source | Generating costs | External costs | Total costs | Rank (least expensive to most expensive) |
|---|---|---|---|---|
| Coal/lignite | 3 | 2–15 | 5–18 | 6/7 |
| Wind | 3–6 | 0.1–0.3 | 3.1–6.4 | 1/2 |
| Natural gas | 5 | 1–4 | 6–9 | 4 |
| Hydropower | 2–8 | 0–1 | 2–9 | 1/2 |
| Geothermal | 5–8 | 0–1? | 5–9? | 3? |
| Biomass | 7–9 | 1–3 | 8–12 | 5 |
| Nuclear | 10–14 | 0.2–0.7 | 10.2–14.7 | 6/7 |
| Photovoltaics | 20–30 | 0.6 | 20.6–30.6 | 8 |

*Source:* Generating costs are from various sources; external costs, except geothermal, are from Sawin 2003.

the original materials are buried thousands of feet below the surface. So although they are still forming today, the rate is imperceptibly slow and existing supplies of oil, gas, and coal are being depleted, oil most rapidly, followed by natural gas and coal.

Petroleum is composed largely of carbon with a bit of hydrogen and some sulfur impurity thrown in. It is the most widely used energy source in the world and America is the biggest consumer. The world in 2003 used 77 million barrels of oil each day (28 billion barrels per year), of which about 20 million barrels per day (7.3 billion per year) are used in the United States (a barrel of oil contains 42 gallons). Oil supplies about 40 percent of the energy we use, with heating, air-conditioning, and transportation the biggest consumers (figure 5.2). One barrel of oil yields 19.4 gallons of gasoline (table 5.2), so any interruption in the supply is felt almost immediately. This was brought home forcefully in 1973 when the Arab countries that then supplied a major portion of our oil decided to turn off the tap because they disapproved of America's foreign policy. Lengthy lines at gas stations developed immediately, emphasizing the nation's vulnerability. Since then we have decreased our dependence on Middle Eastern oil, despite the fact that the share of the oil we use that is

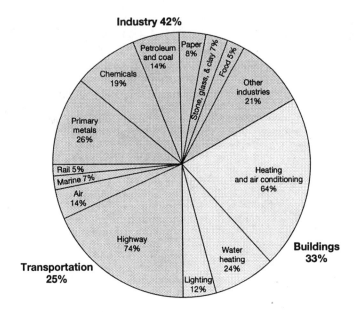

**Figure 5.2**
Energy mix in the United States (Energy Information Administration).

imported has risen steadily since then, from 36 percent to 58 percent. It is projected to rise to 70 percent by 2020.

The world is currently awash in oil but the United States is not. We were the first nation to produce oil (1859), supplied much of the oil used in two world wars, and have used our oil to develop the largest industrial base the world has ever seen. But oil is an exhaustible resource and America is on the downward slope of its supplies. Domestic production has decreased from 9.2 million barrels per day in 1973 to 5.7 million barrels per day in 2003. Further exploration in the contiguous United States or drilling in Alaska cannot reverse this decline for more than six months, according to the U.S. Geological Survey. Our dependence on foreign oil is certain to grow.

### Where Is the World's Oil Located?

Petroleum is found worldwide, although not in every country. Some nations are blessed with amounts far in excess of their needs; other nations have little or none. As was the case with water, the distribution of petro-

Table 5.2
Average product yields from a barrel of crude oil. The total volume of products made is 2.6 gallons greater than the original 42 gallons of crude oil. This represents "product gain," because denser molecules split into less dense ones.

| Product | Gallons per barrel |
|---|---|
| Gasoline | 19.4 |
| Distillate fuel oil (home heating oil and diesel fuel) | 9.7 |
| Kerosene-type jet fuel | 4.3 |
| Coke | 2.0 |
| Residual fuel oil (heavy oils) | 1.9 |
| Liquefied refinery gases | 1.9 |
| Still gas | 1.8 |
| Asphalt and road oil | 1.4 |
| Petrochemical feedstocks | 1.1 |
| Lubricants | 0.5 |
| Kerosene | 0.2 |
| Other | 0.4 |
| TOTAL | 44.6 |

*Source:* American Petroleum Institute.

leum is not fair. Petroleum, like water, occurs in the pore spaces of rocks but at greater depths than potable water. The average depth of oil wells in the United States is 5,000 feet; the deepest producing well depth is 25,000 feet.

As we have all become aware, most of the world's oil is located in the Middle East, the highest proportion in Saudi Arabia, and about four-fifths of the oil is controlled by OPEC (table 5.3). It is a standing joke that the difference between drilling for oil in arid Saudi Arabia and elsewhere is that when you hit water in Saudi Arabia it is considered a success but elsewhere when you drill for oil and hit only water it is considered a failure. The Saudis control about 25 percent of the world's known reserves and, therefore, have a lot of political clout with the major industrial nations, particularly with the world's number one oil user, the United States. The nine nations with the most abundant reserves (table 5.3) can produce (if increasing cost is not a barrier) for at least 50 years at present rates of production. The world is not in immediate danger of running out of oil even though it brings to the surface 37,400 gallons *every second*. In 2000,

Table 5.3
Location of the world's oil reserves as of 2001

| Country | Percent |
| --- | --- |
| 1. Saudi Arabia* | 24.9 |
| 2. Iraq* | 10.7 |
| 3. United Arab Emirates* | 9.3 |
| 4. Kuwait* | 9.2 |
| 5. Iran* | 8.5 |
| 6. Venezuela* | 7.4 |
| 7. Russia | 4.6 |
| 8. United States | 2.9 |
| 9. Libya* | 2.8 |
| 10. Mexico | 2.6 |
| 11. Nigeria,* China | 2.3 |
| 13. Qatar* | 1.4 |
| TOTAL | 86.6 |

*Member of OPEC.
Note: Nations not listed have less than 1 percent (British Petroleum, 2002). If tar sands are included in the table, Venezuela becomes number 1, with Saudi Arabia and Canada close behind, approximately tied for number 2. However, Saudi Arabia has about 1 trillion barrels of oil that may ultimately be recoverable by advances in technology. No country comes close to this.

after a 5-year study, the U.S. Geological Survey raised its previous estimate of the world's crude oil reserves by 20 percent. The oil "shortages" and price fluctuations we experience every so often are political creations and have little or nothing to do with the amount of oil in the ground.

It is worth noting, however, that the U.S. Energy Information Agency expects world oil demand to increase by 54 percent by 2025, largely due to increased demand from developing countries. For example, China's oil consumption grew 6 percent from 2001 to 2002, accounting for all of the world's oil-consumption growth. China replaced Japan as the world's second largest oil consumer. Increases in world oil consumption will have to be met primarily from the oil-rich Middle East, a region with a history of wars, illegal occupations, coups, revolutions, sabotage, terrorism, and oil embargoes. To this list of instabilities may be added growing Islamist movements with various antipathies to the West, particularly the United

**Table 5.4**
U.S. imports of crude oil, 2003, 58 percent of all the oil we used (total = 5.4 billion barrels)

| Country | Percent |
| --- | --- |
| Canada | 16.6 |
| Saudi Arabia | 14.8 |
| Mexico | 13.1 |
| Venezuela | 10.8 |
| Nigeria | 7.2 |
| Iraq | 3.4 |
| Algeria | 3.3 |
| Angola | 3.2 |
| Russia | 2.3 |
| Others (about 20) | 21.6 |
| TOTAL | 100.0 |

*Source:* American Petroleum Institute.

States. Political disruptions in Saudi Arabia that restrict oil exports could cause an oil crisis at any time, maybe tomorrow.

From where does the United States get the oil it imports, about 60 percent of the oil we use? Between one-half and two-thirds of it now comes from countries in little or no danger of revolutionary overthrow (table 5.4). So our present sources of imported oil will probably change little for the forseeable future, with the amounts increasing almost yearly as America's reserves dwindle. Present imports are equal to about 20 super-tankers entering U.S. ports each day. It is striking that since the Arab oil embargo in 1973 that had terrible effects on the American economy, there have been Democratic presidents and Republican presidents, Democratic Congresses and Republican Congresses, yet with the exception of President Jimmy Carter (who wanted to greatly expand nuclear energy) no president or congressional majority has screamed for a change in our major energy source, petroleum. The petroleum industry (and others who support the status quo) is very large, has lots of money to spend, and thus has lots of political clout. Over past decades they have managed to hinder research and development of alternative and non-polluting sources of energy because they have not seen its development as being in their best interests. This attitude seems to be changing.

## Tar Sands

Although we rarely hear about them, two of the largest deposits of oil in the world are located not in Saudi Arabia or Iraq but in Venezuela and Canada. These resources are usually not included in tabulations of recoverable oil because they are not smooth-flowing liquids like crude oil or water but thick, viscous material resembling tar located in the pores of sandstones. They require a special and expensive refining process to be converted into what is termed syncrude (synthetic crude oil). Extracting and refining tar sands require more energy, water, and money than conventional oil, and the refining process produces more greenhouse gases than the refining of conventional oil.

Venezuela contains perhaps 300 billion barrels of recoverable potential syncrude; Canada, between 200 and 300 billion barrels.[4] The total potential of tar sands, therefore, approaches the 600 billion barrels of remaining Middle East reserves that are recoverable with today's technology. At present, only 2 million barrels of tar-sand oil are produced each day at a cost of $13 a barrel, less than half the price of a barrel of conventional oil on world markets. A significant increase in production would require federal government help (from both the United States and Canada) for a multi-billion-dollar investment in research and refining capacity. On the positive side, Canada is our reliable northern neighbor and is politically stable. And Venezuela is likewise far from the ever-volatile Middle East. On the negative side, do we want to prolong our dependence on highly polluting petroleum products? Perhaps more to the point, do we have a choice, given that it will probably require decades for solar and wind power to form the major part of America's energy mix (see below)?

## The Prices of Oil and Gasoline

The prices of crude oil and the gasoline refined from it do not fluctuate wildly, a fact not readily discernible from daily news reports. The price of a barrel of oil in 2000, *adjusted for inflation*, has remained stable since the late 1800s, quite a small variation over so long a period (figure 5.3). The only exception was the price spike related to the Iran–Iraq War in the 1980s. And the average cost of gasoline at the pump (adjusted for inflation) has decreased fairly continuously, although it is hard for the average American to believe (figure 5.4). However, the stable price of oil (in real dollars) and declining

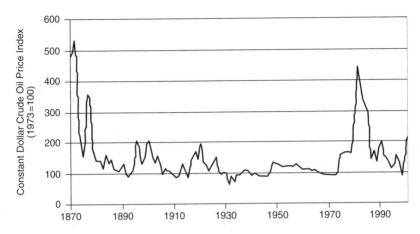

Figure 5.3
Price of a barrel of oil, adjusted for inflation. 100 = $10/barrel. 200 = 20/barrel, and so on (Department of Energy).

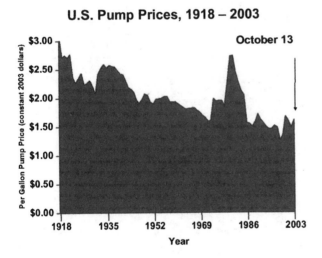

Figure 5.4
Price of gasoline at the pump, adjusted for inflation (2003 dollars, American Petroleum Institute).

price of gasoline (in real dollars) may be coming to an end because of the rapidly growing demand from China and India.

The cost of gasoline consists of three parts, each of which can vary yearly. According to the Energy Information Administration, in March 2003 the relative costs in the United States were: price of crude oil, 46 percent; refining, 19 percent; marketing and distribution, 11 percent; and federal and state taxes, 24 percent. The state/federal taxes per gallon vary widely among the states, from a low of 43.2 cents in Georgia to a high of 85.4 cents in California.[5] Most of the higher cost of gasoline in Europe and elsewhere (table 5.5) consists of taxes, not the cost of the crude oil, refining, and marketing.

The typical $1.50-per-gallon cost of gasoline in 2003 and its apparent constancy for the past hundred years is extremely deceptive. Hidden costs—from U.S. tax breaks to oil corporations to armed securing of supply lines and sea routes worldwide to indirect costs such as the Iraq

Table 5.5
International gasoline prices, March 2000. European nations place heavy taxes on gasoline. Major producing nations have low prices.

| City | Price (U.S. dollars) |
| --- | --- |
| Hong Kong | 5.24 |
| London | 4.83 |
| Amsterdam | 4.48 |
| Oslo | 4.38 |
| Espoo, Finland | 4.35 |
| Tokyo | 4.34 |
| Copenhagen | 4.29 |
| Paris | 4.22 |
| Seoul | 4.22 |
| Stockholm | 4.01 |
| United States | 1.43 |
| Abu Dhabi, United Arab Emirates | 0.98 |
| Riyadh, Saudi Arabia | 0.93 |
| Manama, Bahrain | 0.82 |
| Kuwait City, Kuwait | 0.76 |
| Jakarta, Indonesia | 0.52 |

Source: Runzheimer International.

wars—boost the real gallon price to something like \$12–\$20.[6] Does anyone really believe that our "humanitarian concern" for the people of Kuwait, Saudi Arabia, and Iraq would have caused the U.S. to go to war if the oppressed and invaded were citizens of Myanmar, Uganda, Cambodia, or Rwanda? Recent history suggests otherwise.

### Will the World Run Out of Oil?

The original amount of potentially recoverable oil in the earth's crust is estimated to have been about 2 trillion barrels (not counting tar sands), of which 1 trillion, half the original amount, has already been produced. And oil demand is rising rapidly as large nations such as China and India industrialize. The discovery of new oil fields has declined despite significant improvements in exploration science and technology. Oil discoveries have declined steadily from about 470 billion barrels between 1950 and 1960 to 110 billion barrels between 1990 and 2000.[7] Rather than increasing discoveries, better technology has mostly served to reveal that there aren't many new oil fields left to be found.

Oil production tends to accelerate until about half the oil is drained from oil fields; from that point onward, oil is produced more slowly and more expensively. No one disputes that there will be serious repercussions as global oil production peaks and begins to decline. The federal government and 44 states now offer financial incentives to encourage industry and homeowners to increase their use of alternative energy sources. And manufacturers of major electrical appliances such as refrigerators, washing machines, and water heaters have increased the efficiencies of their products.

### The Future of Petroleum

The use of petroleum as a source of energy is so ingrained in America's, and the world's, industrial economies that doing away with it completely is unthinkable for the forseeable future, despite looming price increases as world supplies dwindle. But a significant reduction in its use is certainly within reach during the next few decades. Many alternative, inexhaustible, nonpolluting sources of energy are already being used, although in relatively small amounts, and new ones are being developed. Several major petroleum companies have seen the handwriting on the wall and have made major investments in such things as solar panels, wind turbines, and fuel cells. British Petroleum (BP) is now the leading manufacturer of solar cells. Shell is pioneering research in fuel-cell technology. Because of concerns about pollution, the use of oil as a source of energy will probably decline over the coming decades. As Sheikh Zaki Yamani, former oil minister of Saudi Arabia, put it, "The stone age did not end for lack of stones, and the oil age will end long before the world runs out of oil." It will end because of concerns about rising prices, national security, and pollution.

**Natural Gas**
Natural gas is commonly found associated with and is produced with crude oil deposits. It also occurs as deposits not associated with oil and is the second most abundant energy resource after coal. It supplies 24 percent of America's energy needs. The rule of thumb is that in terms of energy yield a barrel of oil is equivalent to about 5,500 cubic feet of gas. Prices for gas are quoted in units of 1,000 cubic feet and the unit price in April 2003 was about $5.50, up from $3.65 a year earlier and $2 four years earlier. Residential gas prices increased 40 percent between 1995 and 2002.[8] Russia has by far the largest reserves (table 5.6). It is the Saudi Arabia of natural gas. As it comes out of the ground the gas is usually 90–95 percent methane, the same gas that seeps out of landfills (chapter 3). Natural gas is abundant. At present rates of consumption, North America's reserves are sufficient for the next hundred years. However, the use of natural gas is increasing rapidly and the reserve estimate of 100 years may prove overly optimistic. We cannot be sure because new deposits of natural gas will probably be found between now and then. Exploration continues.

The United States uses 27 percent of the world's production. Natural gas is cheaper than petroleum for the same amount of energy obtained, and burning it produces 30–60 percent less carbon dioxide than the same amount of energy derived from oil. For these reasons 80 percent of new power plants are gas-fired. Gas currently generates 18 percent of our electricity. The Energy Information Administration predicts that the use of natural gas in the United States will grow by 54 percent by 2025, while production will rise only 35 percent. Imports are expected to supply 22

Table 5.6
Where the world's proven reserves of natural gas are located, 2001. All other nations have less than 2 percent.

| Country | Percent |
| --- | --- |
| 1. Russia | 30.7 |
| 2. Iran | 14.8 |
| 3. Qatar | 9.3 |
| 4. Saudi Arabia | 4.0 |
| 5. United Arab Emirates | 3.9 |
| 6. United States | 3.2 |
| 7. Algeria | 2.9 |
| 8. Venezuela | 2.7 |
| 9. Nigeria | 2.3 |
| 10. Iraq | 2.0 |
| TOTAL | 75.8 |

*Source:* British Petroleum, 2002.

percent of our needs in 2025. At present about 20 percent of our natural gas is imported, 95 percent of it from Canada.

As we saw earlier, increased oil imports will have to come from the politically unstable Middle East. Where will the projected increase in imports of natural gas come from? Our geographic neighbors Canada and Mexico are not in the top ten in terms of world gas reserves (table 5.6) and Canada, like the United States, is increasing its domestic use. The apparent choices are Russia and the ever-volatile Middle East. But this poses an expensive transportation problem. For economic reasons, natural gas is cooled to a liquid to be transported in ships—as liquefied natural gas or LNG. (This process frees up cargo space; liquefying the gas is like shrinking a gallon to a teaspoon.) Natural gas (methane) liquefies at -258°F. It must be cooled at the shipping terminal, pumped into heavily insulated cargo holds, and then heated as it is discharged at the receiving terminal. America's imports of LNG have risen from 20 billion cubic feet in 1995 to 85 billion cubic feet in 1998 to 230 billion cubic feet in 2002.[9] New shipping terminals for receiving LNG are now being built on our coasts as fast as construction materials can be trucked there.

Motor vehicles can be run on compressed natural gas (CNG). Converting a car to CNG is easy because, like gasoline, it burns well in spark-ignition engines. Within the past 10 years some cities have converted their municipal vehicles from gasoline to CNG. The vehicles are refueled at central municipal locations. Very few service stations for the general public provide CNG, however. One disadvantage of using CNG in cars is that the fuel tank is heavy and bulky. For Americans, who live in a large country and drive longer distances than people in other nations, this is a distinct disadvantage.

**Natural Gas Hydrates**
During the past decade there has developed a great interest in widespread deposits of methane trapped in icelike crystals that are abundant under the shallow ocean floor. These methane hydrate accumulations look like normal ice but burn if touched by a flame. Estimates of the extent of these strange deposits indicate they could contain more than twice as much methane as known conventional gas reserves. In December 2003, scientists announced that they had solved the technological problem of producing natural gas from hydrate. Commercial production may be on the energy horizon.

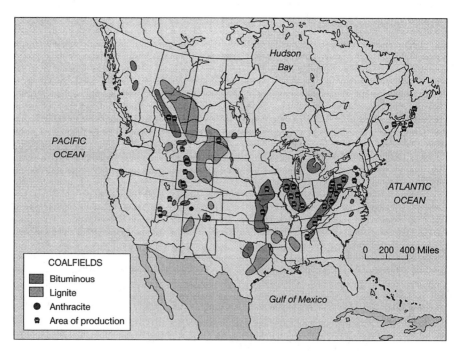

**Figure 5.5**
Coal deposits in North America (U.S. Geological Survey).

Table 5.7
Where the world's proven reserves of coal are located, 2001. All other nations have less than 1 percent.

| Country | Percent |
| --- | --- |
| 1. United States | 25.4 |
| 2. Russia | 15.9 |
| 3. China | 11.6 |
| 4. India | 8.6 |
| 5. Australia | 8.3 |
| 6. Germany | 6.7 |
| 7. South Africa | 5.0 |
| 8. Ukraine, Kazakhstan | 3.5 |
| 10. Poland | 2.3 |
| 11. Brazil | 1.2 |
| TOTAL | 88.5 |

*Source:* British Petroleum, 2002.

## Coal

World coal reserves are estimated to be 1,000 years and the United States has the world's largest reserves (figure 5.5, table 5.7). Coal supplies about the same percentage of America's energy as natural gas. But coal is essentially pure carbon, and as a result coal combustion emits 25 percent more carbon dioxide per unit of energy obtained than oil (natural gas is cleaner than either coal or oil). Coal is the single biggest air polluter in the United States. It produces seven times more soot and 50 times more sulfur dioxide than natural gas.[10] The United States is the biggest user of coal (26 percent of world use). The Department of Energy predicts only 1 percent annual growth in coal use through 2015. Because of the amount of coal and oil we burn, we are the largest emitters of heat-trapping carbon dioxide gas, a major producer of soot, and we produce 21 percent of the world's carbon dioxide emissions.

Fifty-two percent of America's electricity is generated by the burning of coal,[11] and the dominance of coal is expected to continue for several decades. Electricity generated by coal-fired plants costs half as much as electricity from oil- or gas-fired plants. At one time nearly all power plants were coal-fired, but because of concerns about pollution ("external costs") few new coal-fired plants are being built in the United States. Old plants should be gradually retired as new gas-fired plants are constructed. Natural gas should be the fuel of choice for new power plants because of its relatively low external costs (table 5.1). But the price of natural gas is increasing rapidly, which will likely slow the retirement of the coal burners. At present, electricity generation accounts for 35 percent of our emissions of carbon dioxide and nearly 40 percent of the nitrous oxide, important contributors to enhanced global warming.

China and India are the world's most populous countries and are undergoing rapid industrial development. They have coal reserves sufficient to last for centuries and are using it to fuel their industrial growth. China plans to almost triple the capacity of its coal-fired power stations by 2020. Coal supplies about 66 percent of China's total energy needs. In 2002, coal consumption in China increased a massive 28 percent and increased an additional 10 percent in 2003. China's energy use accounted for more than two-thirds of the increase in global energy consumption.[12] China's emissions of greenhouse gases between 2000 and 2003 almost equalled those of all industrialized nations combined.

Developing countries do not have the tight emission controls on coal combustion that exist in the United States. The air pollution in China's major cities is already disastrous. People walk around in Beijing wearing surgical masks to filter out the soot. In 2000, China publicly stated its intention to greatly increase its reliance on natural gas and to "increase significantly" its use of renewable energy sources.

## Alternative Energy Sources

The best way to predict the future is to invent it.
—Alan Kay, computer pioneer

Renewable energy sources are in the position of petroleum about a century ago, accounting for only a tiny part of the current energy picture but poised for a dramatic increase. Shell predicts that renewables will become fully competitive with conventional sources by 2020, and by 2050 they predict that renewables will dominate the primary energy market.[13] Other scenarios by research groups envision only a 40 percent or 25 percent niche for renewables, but all see a rapid increase in their use during the next few decades. In 1999 the International Energy Agency noted that "the world is in the early stages of an inevitable transition to a sustainable energy system that will be largely dependent on renewable resources."[14]

Renewable sources of energy are those that are not fossil fuels, are inexhaustible, and produce little or no pollution. They supply 6 percent of America's energy. They include biomass (47 percent), hydropower (45 percent), geothermal power (5 percent), wind power (2 percent), and solar power (1 percent).[15] A significant minority of Americans would add nuclear power to this list as well.

Renewable energy sources have historically had a hard time breaking into markets that have long been dominated by fossil-fuel systems. This is partly because the new technologies are only now starting to be mass produced and have had high capital costs relative to established systems, but also because coal-, oil-, and gas-powered systems have benefited from many federal subsidies over the years. These include military expenditures to protect oil exploration and production interests overseas, the costs of railway construction to enable economical delivery of coal to power plants, depletion allowances to oil companies, and a wide range of tax breaks.

**Nuclear Power**

Nuclear power is difficult to categorize although the source of the power, enriched uranium, is abundant. The world's resources of uranium are sufficient for only about 50 years at present prices, but a doubling in price would create a tenfold increase in economically minable reserves[16] without seriously affecting the cost of uranium-generated electricity. The United States contains only 3 percent of the world's known reserves (table 5.8).

Early in its development about 50 years ago, nuclear power was widely viewed as the ultimate panacea for America's energy problems, a safe, inexhaustible, and pollution-free source of energy. One pound of uranium can yield as much energy as three million pounds of coal. But now, there are heightened concerns about plant safety and pollution from radiation. No nuclear plants have been ordered in the United States in almost 30 years despite our continually increasing energy needs.

Table 5.8
Known recoverable resources of uranium

| Country | Percent |
| --- | --- |
| Australia | 28 |
| Kazakhstan | 15 |
| Canada | 14 |
| South Africa | 10 |
| Namibia | 8 |
| Brazil | 6 |
| Russia | 4 |
| United States | 3 |
| Uzbekistan | 3 |

Source: World Nuclear Association, www.world-nuclear.org/info/inf75.htm.

Because of increased costs mandated by increased emphasis on safety, perpetual lawsuits by environmental groups seeking to halt construction, and the well-founded fears about radiation, nuclear power is unlikely to be the dominant player in America's energy mix that was envisioned 50 years ago. Nuclear power currently supplies 8–9 percent of America's energy needs and 20 percent of its electricity. But electricity generated by nuclear power is twice as expensive as electricity generated by a coal- or gas-fired plant (table 5.1).

The nuclear power industry, now about 50 years old, has peaked and has begun an economic meltdown. The last U.S. reactor under construction was authorized in 1979 and completed in 1996. In 1974 there were 224 nuclear plants in operation or on order in the United States; today there are none on order and 103 in operation at 65 plant sites in 31 states, more than in any other country. Their average age is 20 years.

The United States has nearly one-fourth of the world's nuclear plants. Many existing plants are on the verge of closing because of lawsuits by environmental groups, high operating costs, or obsolescence. Ten percent of the nation's aging reactors have been removed from service in the last ten years. Between September 2000 and April 2001 there were at least eight forced shutdowns due to equipment failures caused by aging. In 2002, 48 of 59 nozzles in the reactor vessel head of a reactor in Virginia were found to be cracked. Yet tests the previous year reported that none of the nozzles were cracked.[17] As much as 40 percent of the U.S. nuclear capacity is vulnerable to permanent shutdown due to high costs. Power plants that cost a billion dollars or more to build have been sold for less than $100 million. The Department of Energy estimates that 31 percent of the country's nuclear capacity will be closed by 2015. The venerable U.S. engineering firm, Stone & Webster, which had a hand in building most American plants, filed for bankruptcy in 2000. Operational dangers at America's nuclear facilities will not end until the last nuclear power plant is closed down and dismantled many decades from now.

The industrialized nations of the world have begun to agree with the concerns of the American public. At the end of 2002 there were 437 nuclear reactors operating worldwide. They produce 17 percent of the world's electricity, similar to the percentage in the United States. France is the leader in nuclear power use and obtains 77 percent of its electricity from its 59 reactors. But even they are having second thoughts about their overwhelming dependence on the nuclear option. They have not built any new reactors since the Chernobyl disaster. And in a remarkable reversal of governmental policy, Germany in 2000 became the first industrial power to announce its abandonment of nuclear power. All of its 19 nuclear reactors will be shut down by 2020. At present, Germany gets 30 percent of its energy from this source. As part of its conversion program, Germany has become the world wind-energy leader, with the wind now supplying 4.5 percent of its electricity needs. The Germans now have 38 percent of global wind-energy installations.[18] Sweden, Austria, Belgium, Denmark, and Italy have also renounced nuclear power. Most other nations in the European Union have no plans to build new plants when their current reactors reach the end of their operational lifetime of about 40 years, with the exception of frigid Finland, which currently gets 33 percent of its electricity from nuclear power. The parliament has approved plans to build a fifth reactor in the country.

Some of the newer and pending members of the EU in Eastern Europe, though, have plunged ahead with plans for nuclear power. Plants are under construction in Slovakia, Romania, Ukraine, and Russia and proposals are on the table in Bulgaria and Lithuania. The Romanian government wants to quadruple nuclear power's share of the country's total power output, and sell excess power to its neighbors.

While no nuclear plants have been built in the United States in recent years, existing facilities have substantially improved their performance and lowered operating costs. Further, it has become common practice to request extension of the operating licenses of nuclear plants from the U.S. Nuclear Regulatory Commission.[19] The first extension, for 20 years, was granted in March 2000. As a result, the downturn in nuclear generating capacity and generation previously expected is now anticipated to be delayed or eliminated. A more recent phenomenon has been the upgrading of nuclear plant capacity. And President George W. Bush has proposed a massive increase in nuclear power development, aided by government subsidy of half the cost of building nuclear plants. The Department of Energy predicts an increase in nuclear capacity and generation. Chapter 9 presents an extended discussion of radiation and the disposal of our ever-growing mountain of nuclear waste.

### Nuclear Power and Terrorism

The explosion at Chernobyl in 1986 resulted from human error. But following the attacks in New York and Washington by terrorists on September 11, 2001, concerns have arisen about the possibility of similar attacks on America's nuclear power installations. None of these facilities were built to withstand the impact of a fuel-laden 757 or 767 jetliner, but expert opinion is that reactor containment vessels could withstand even this impact, although uranium reprocessing facilities could not.[20] There are no uranium reprocessing facilities in the United States, but there are several in Europe and Asia. The impact of a fuel-laden 757 or 767 could not cause a nuclear explosion but could release radiation sufficient to cause the permanent evacuation of hundreds of square miles of densely populated suburbs and cities. The accident at Chernobyl rendered more than 1,000 square miles unsafe for human life for at least 100 years.

France has installed surface-to-air missiles around its nuclear reprocessing complex at La Hague. In Britain, armed guards and concrete blocks have been installed at entrances, and exclusion zones up to 2,000 feet high within a 1.8-mile radius of 12 nuclear power stations are now in place. At present, the United States has no comprehensive plan to protect its nuclear plants from terrorist attack.

Recent evidence suggests that plane hijackings are unnecessary for terrorists to successfully attack a nuclear power plant. In January 2003, 19 Greenpeace activists stormed the U.K.'s Sizewell power plant, avoiding the guarded entrances and scaling the reactor without resistance. Their goal was only to expose the plant's vulnerability, but if the intruders had been actual terrorists the result would have been catastrophic.

## Hydropower

Hydropower is the leading renewable energy source used by electric utilities to generate electric power. Most of the nonnuclear alternative or renewable energy in the United States comes from hydropower. The United States is the world's largest producer of electricity by hydropower; it produces 7 percent of our needs but is unlikely to grow significantly in the future. The reasons are different but easily understood. Hydropower generation requires the damming of rivers to impound large reservoirs of water that can be released as needed to spin the turbines of the hydroelectric plant in the dam (figure 5.6). Such dams are abundant along our country's rivers, particularly in the Pacific and Rocky Mountain states, where 70 percent of the hydroelectric power in the United States is generated. Most of the best sites for dam construction are already being used and Mother Nature is not creating any significant new rivers.

From an environmental viewpoint, hydroplants have several advantages over energy production using oil, natural gas, coal, or nuclear energy. They release no radiation, have low operating and maintenance costs, produce no particulate air pollution, and produce no sulfur dioxide or nitrogen oxides. Further, hydroplants last two to ten times longer than power plants that use fossil fuels. The drawbacks to hydropower stem from the massive disruption the dams cause in the natural regime of the river and its surroundings. Large areas behind the dams are permanently flooded, causing major ecological changes along the river's course. And recent studies indicate that the decay of organic matter continuously washed into a reservoir from upstream vegetation generates a great deal of methane, a potent greenhouse gas. Also, the reservoir and the dam at its far end block the sand and mud that used to move downstream to the ocean. So the reservoir must

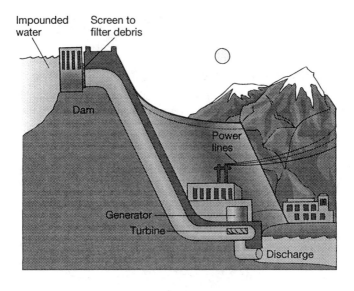

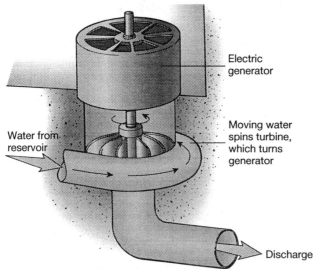

Figure 5.6
Cutaway of a typical reservoir and hydroelectric plant, with detail of power generation (D. Kash et al., *Energy Alternatives,* Report to the President's Council on Environmental Quality, 1975).

be dredged at great expense to maintain peak power generation at the dam site. In addition, restricting the flow of water downstream of the dam changes the natural pattern of erosion and sediment deposition, and this commonly has a negative impact on downstream communities.

## Geothermal Power

Geothermal energy is the natural heat of the earth's interior. Sites where the earth's heat can be tapped most effectively are identified by the presence of steam vents, an indication that it is very hot very near the surface. The cause of the steam is the heating of groundwater by molten rock from the earth's interior that has risen to within a few thousand feet of the surface. This happens in volcanic areas around the world. The nation of Iceland, composed of a group of islands that are active volcanic peaks in the North Atlantic, uses geothermal heat for much of its total energy supply. About 40 countries extract energy from geothermal sites. In the United States, geothermal sources supply less than half of 1 percent of the energy consumed. The major facility is at The Geysers in northern California, an important regional energy source that has been operating since 1960.

In a geothermal steam power plant, steam emanating from the vents is routed directly to a turbine that drives an electric generator, eliminating the need for the boilers and fossil fuel of conventional power plants. The Geysers uses steam generation and delivers electricity at 4–5 cents/kWh.[21] It supplies 7 percent of California's energy.

However, hot-water resources are much more common than steam resources. Hot-water plants are operating in California, Nevada, Utah, and Hawaii. In these cases, where the water from the vents is hotter than 400°F (confining pressure keeps it from boiling), the water is sprayed into a tank held at much lower pressure than the fluid, causing some of the water to flash to steam. The steam drives a turbine that drives a generator. Hydrothermal power generated from hot water is less efficient than steam power and hence more expensive.

Geographic constraints are an important consideration for hydrothermal power generation. All of the known high-temperature hydrothermal reservoirs, whether water- or steam-dominated, are located in Western states. For direct heat applications, in particular, it is crucial that the heat source be located near the demand, because piping costs and energy losses increase rapidly with distance.

## Bioenergy

Bioenergy is energy contained in natural biological materials such as plants and animal waste. Plants replenish themselves and animals never stop producing waste, so that biomass is inexhaustable. It includes forests, grass, straw, cut wood, waste paper, agricultural plant residue such as corn husks and cobs, manure, and gases and liquids obtained from these sources. If biomass is already dead, burning it simply speeds up the natural decay of dead plants, which is environmentally benign. Killing growing trees, however, leads to deforestation and increased soil erosion, and slows the rate of removal of carbon dioxide from the atmosphere because it decreases photosynthetic activity.

Most biomass is burned for heat rather than for generating electricity. Biomass can supplement about 10 percent of the coal in a coal-fired power plant or can be burned alone in a conventional boiler. It can also be digested in an oxygenless environment to produce methane that can then be used as a power source. Solid biomass can be converted into liquid fuels that power cars or industrial operations. Ethanol, used as a gasoline additive, is produced by fermentation. Biodiesel is produced by combining alcohol with oil extracted from soybeans, rapeseed, animal fats, or other biomass. Biomass supplied 3 percent of America's energy needs in 2002. Most biomass power-generating facilities are located along the East and West coasts.

## Solar Power

Solar energy currently supplies 1 percent of America's energy needs but has great potential for growth (figure 5.7). Solar cells, also called photovoltaic cells, convert solar energy directly into electricity. Most solar cells are thin wafers composed of purified crystals of silicon with trace amounts of cadmium and gallium added. When sunlight hits these wafers they emit electrons, which flow as an electric current. Unfortunately, current commercial solar-cell wafers have a maximum efficiency of only 15 percent. Hence, solar power is much more expensive than wind power.

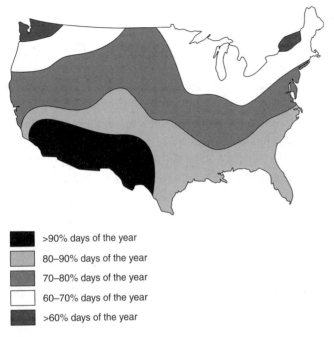

■ >90% days of the year

▢ 80–90% days of the year

▨ 70–80% days of the year

▯ 60–70% days of the year

▧ >60% days of the year

**Figure 5.7**
Solar energy in the United States (Department of Energy and National Wildlife Federation).

A single wafer of silicon emits only enough energy to light a flashlight battery, so many cells must be wired together in a solar panel to produce even 30–100 watts of electric power. Growth in world production of solar cells in 2001 was 36 percent, marking the fourth straight year of growth of at least 30 percent.[22] In 2001 Japan produced 43 percent of the world's solar cells, with the United States second at 25 percent. U.S. production of photovoltaic cells expanded by 11 percent in 2001 and for the first time in 10 years most shipments were intended for domestic installations, an encouraging sign of America's use of alternative energy. However, solar power still generates only a minuscule 0.02 percent of our electricity. Assuming 20 percent efficiency for solar cells, it would require an array of solar panels the size of the combined areas of Connecticut and Rhode Island to supply all of America's current electrical demand.

Solar panels have been developed that can replace normal roof shingles on homes. In California, hit hard by soaring electricity prices and rolling blackouts, the Los Angeles Department of Water and Power in 2000 established a $75 million budget to support as many as 100,000 solar rooftop installations over the succeeding 5 years. According to British Petroleum, it costs between $10,000 and $40,000 for a solar roof that lasts 25 years that can provide enough energy to meet the requirements of a whole house. However, with federal and state rebates and incentives the cost can be reduced by as much as 60 percent. According to Fannie Mae the average home energy bill in America is $1,331 per month, so if we generously assume that with rebates it would cost $20,000 to totally convert your house to solar energy, the cost would be recouped in 15 years. Starting in year 16 you would save $1,331 (at least) every year. On the negative side, Americans change domiciles every 7 years, on average, so that someone other than you stands a good chance of being a major beneficiary of your investment. As noted by the director of the federal government's National Renewable Energy Laboratory, "Solar is sexy and everybody loves it, but the fact is it remains too costly" for anything but niche applications.

In their new gas stations, petroleum giant BP installs transparent solar panels located atop a transparent canopy that shelters motorists and generates 15 percent of the station's power. More than 250 BP gas stations are now running partly on sunshine.[23] Unfortunately, there are few BP gas stations in the United States. BP Solar is the world's largest manufacturer of solar technology, as well as one of the largest users of solar power.

In 2000, Alan Heeger received a Nobel Prize for his development of electrically conductive plastics. They absorb light and produce electricity. When used in solar cells instead of purified silicon they will greatly reduce the cost of the cells and make them competitive in price with major power sources that run on fossil fuels. At present, the plastics are too inefficient to be commercially useful, converting only a few percent of the sun's rays into electricity, but researchers are optimistic they can match the 15 percent efficiency of silicon crystals within a few years.[24]

In July 2003, a subsidiary of the Boeing Company achieved a record conversion efficiency of 36.9 percent for solar cells.[25] The subsidiary of Boeing that produces the PV cells believes efficiencies greater than 40 percent are possible. Several of these solar modules are already being tested throughout the world.

nd. And it is worth noting that wind is immune to wild price fluctua-
ns, unlike oil and natural gas.

The cost of wind energy is strongly affected by average wind speed and
e size of a wind farm. Because the energy that the wind contains is a
nction of the cube of its speed, small differences in average winds from
e to site mean large differences in production and, therefore, in cost. The
me wind plant will generate electricity at a cost of 4.8 cents/kWh in 16
ph winds, 3.6 cents/kWh in 18 mph winds, and 2.6 cents/kWh in 20.8
ph winds. Larger wind farms provide economies of scale. A 3-megawatt
nd plant generating electricity at 5.9 cents per kWh would generate
ectricity at 3.6 cents per kWh if it were 51 megawatts in size.

The major disadvantage of wind power is that the wind is intermittent,
 that turbine operators can't tell the local utility how much power they
n reliably provide. Windmills also can't be readily ramped up like a
ower plant to meet varying demand. Better ways are needed to store
ind energy, and fuel cells now being developed are one possibility.

## ydrogen Power

*ydrogen power* refers to using a combination of hydrogen and oxygen
 produce electricity, with water and heat as by-products. The electricity
owers the vehicle's motor and charges its battery. The oxygen-hydrogen
action is done in a fuel cell, which is similar to a battery in design but
oes not run down or need recharging. Fuel cells are clean and quiet, and
ey have few moving parts and produce little noise or pollution. They
romise to be one of the most important power sources of the twenty-first
ntury. Unfortunately, they are still quite pricey. Though costs have
llen, you still have to pay about $5,000 per kilowatt for their electric-
y.[29] Car engines do the same thing for $50. But fuel cells are currently in
 explosive phase of research and development, and there is little doubt
eir cost will continue to drop. And they are certain to become cheaper
 the economies of mass production kick in.

Fuel-cell technology has the potential to revolutionize electricity gener-
tion. First, they'll replace batteries, which are very expensive for the
nall amount of power they deliver. Methanol is likely to be the fuel of
hoice as a source of hydrogen. Toshiba plans to sell a laptop computer
owered by a fuel cell soon. Within a few years, fuel cells will provide heat

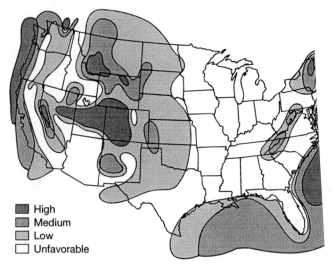

**Figure 5.8**
Map of wind resources (U.S. Department of Energy, Wind Energy
of the United States, 1987). Interestingly, Chicago, long nicknam
City," lacks the prevailing winds necessary to produce useful ene

**High**
**Medium**
**Low**
**Unfavorable**

## Wind Power

Wind power is tied with geothermal energy and second o
power as the least expensive nonpolluting way to genera
Many places in the United States are windy (figure 5.8). Th
able areas are along our coastlines because ocean winds are
steadier. Several wind farms along the northeastern coast ar
the planning stages. Nevertheless, despite rapid growth in
wind power still provides only 0.7 percent of the nation's
even though federal government researchers have shown tha
is feasible with current technology. California leads the v
power development with 52 percent of America's usage; Te:
at 18 percent.[27] Wind capacity in the United States grew a
rate of 22.3 percent between 1999 and 2003. Among onsho
America's vast, flat, and windy Great Plains has the potentia
America's center of wind power. Some Midwestern farmers
that they can generate more income per acre from the electi
ated by a wind turbine than from their crop or ranching
Wind strength could become a major determinant of the futu

and electricity for homes and offices. The source of the hydrogen will probably be natural gas. Home fuel cells would decentralize the production and delivery of electricity. It would eliminate losses incurred as the current travels hundreds of miles over power lines from where it is produced to where it is used. It would also be cheaper. It is three to five times less expensive to ship energy as a gas in a pipeline than as electricity in a high-voltage transmission line.[30] But the most visible and exciting development is just arriving; fuel-cell powered buses, trucks, and cars are starting to purr onto the streets of our cities, despite their slightly higher purchase price.

Fuel cells are attacking a giant. The internal-combustion engine may be the most entrenched technology in existence—tooled and retooled over 150 years to reach the limits of performance and reliability, manufactured in enormous quantities, and supported by a ubiquitous refueling and repair infrastructure. Fuel cells are the David against the internal-combustion-engine Goliath. But unlike the uncertainty that surrounded that epic battle 3,000 years ago, there is no uncertainty about the current struggle. Fuel cells will win.

The vanguard nation in the transition to a hydrogen-energy economy is Iceland, where in 1999 the government officially installed a policy to produce hydrogen using their readily available renewable resources, geothermal and hydroelectric power. In April 2003 they opened a hydrogen fuel station. It will fuel buses in Reykjavík, the nation's capital. Back in 1874, Jules Verne argued in *The Mysterious Island* that when fossil fuels run out, hydrogen "will furnish an inexhaustible source of heat and light." It seems that another piece of science fiction is about to come true.

A key environmental question about the use of fuel cells is the source of the "raw materials," oxygen and hydrogen, needed to generate the electricity. The oxygen can come from the air, which contains 21 percent oxygen. No problem there. The difficult question is the source of the hydrogen. There is none in the air, you can't mine it, and hydrogen gas is difficult to store. The first generation of fuel cells in cars or homes will likely obtain its hydrogen from a fossil fuel, either gasoline or natural gas. Both contain lots of hydrogen, but the use of gasoline as a source of hydrogen yields an emission reduction of just 22 percent compared to an internal-combustion engine. The reduction is 72 percent if natural gas is used.[31] According to the Department of Energy, 96 percent of hydrogen produced in the world today comes from fossil fuels, mostly natural gas.

Maximum environmental benefits from fuel cells will only happen when it becomes economically feasible to extract the hydrogen from water rather from a hydrocarbon such as gasoline or natural gas, using wind or solar energy to generate the needed electricity, a prospect that is decades in the future. The need to free our energy supply from carbon is a worthy and necessary goal if we are ever going to solve many of our water- and air-pollution problems.

However, even when hydrogen generated by electrolysis of water becomes a reality, it is sobering to keep in mind the response given by a fourth-grader when asked what environmental effects cars would have if they were powered by a nonpolluting source of energy such as fuel cells: "You would still have to cut down the trees and pave everything over for roads." And cars will still sit in traffic jams and average a lower effective speed than bicycles. Fuel cells can reduce the pollution generated by motor vehicles but will not change the fact that America needs to switch from private to public transportation.

### Oceanic Energy

A more futuristic source of power to generate electricity involves harnessing the oceans that cover 71 percent of the earth's surface. The potential is great but so are the technological obstacles. In coastal areas with a large tidal range, flowing tidal waters contain large amounts of potential energy, but with existing technology it is only practical to exploit coastal sites where the tidal range exceeds 16 feet. There are perhaps two dozen such sites around the world, only one of which is in the United States, in Maine. One of the main barriers to the increased use of tidal energy is the cost of building tidal generating stations. For this reason, there is only one major power generating station in operation in the world, at the mouth of the La Rance River estuary on the northern coast of France. It has been operating since 1966.

Wave energy can also be harnessed as a power source and one commercial facility came online off the coast of Scotland in 2000. Electricity is generated at a cost of 7 cents per kWh, a price not yet competitive with existing power sources. The best wave-energy regions tend to be on seacoasts at the receiving end of waves driven by the wind over long stretches of water. As the waves travel, the winds continually pump energy into them. By the time the waves hit the coast, they're brimming with power. In the United States, the greatest wave energy potential is on the coasts of Washington, Oregon, and California, but no development is currently in the works. The Department of Energy is not sponsoring wave-power research.

Another possible source of power takes advantage of the temperature difference between the relatively warm ocean surface temperatures and the colder temperatures at depth. Most of the research into ocean thermal energy conversion (OTEC) has taken place in Hawaii, where a small plant was tested in 1999. The plant is not currently producing electricity.

For an OTEC plant to function there must be an average monthly temperature difference of at least 40°F between the ocean surface and the temperature at depth. In an OTEC plant, warm surface water is pumped in to heat liquid ammonia. The liquid ammonia boils at -29°F, so it readily vaporizes. The vapor drives a turbine to generate electricity. The vapor is then cooled by cold water pumped from depths of 2,000 feet or deeper. Cooling the ammonia condenses it back into a liquid, so it is ready for the next cycle. The only pollution produced is carbon dioxide if a hydrocarbon fuel is used to operate the ammonia compressor.

Unfortunately, OTEC is expensive. One-third of the energy produced is needed to pump the deep cold water to the surface. Construction costs are very high, and the ocean environment is hostile. Storms, hurricanes, saltwater corrosion, and fouling by algae and barnacles add to postconstruction expense.

## Conclusion

Many methods are available to generate the energy America needs to run its economic engine. At present the vast majority of this energy is produced by oil, natural gas, and coal, fuels rich in carbon and, therefore, rich in pollutants as well. This needs to change. But change will not happen overnight because of the enormous existing infrastructure for producing and processing these fuels, and the high cost of conversion to cleaner and more efficient energy sources. Oil and coal are a clear and present danger to human health and the environment and should be phased out, a process that will take at least several generations. Natural gas is abundant and less polluting than oil and coal, and its use is likely to increase over the next few decades.

Hydropower, geothermal energy, and biomass can never be our principal energy sources because of natural limitations on their abundance. The way to go appears to be research into lowering the cost of solar power and continuing the increased use of wind power, already the cheapest renewable, nonpolluting source of power available. Wind power fed into existing electricity grids and used to power fuel cells appears to be the best option at present as an energy source, and this pairing is now becoming visible on the horizon. Fuel cells are entering the market in cars, buses, and computers, but the fuel cells will initially be powered by hydrogen obtained from hydrocarbons of one type or another. This will change over time, but the massive use of wind energy appears to be at least a decade away.

The difficulty of reforms in energy usage is that the American political process produces laws, not long-range policies. The emphasis is on temporary or short-term palliatives, solutions that will stem or at least slow the tide of pollution for a while, at least until the next election. Each law reflects the political pressures of the moment. And once a law is passed, it is difficult to repeal or modify significantly until its long-term harmful consequences reach crisis proportions. It is only a slight exaggeration to say that we have an ambulance form of government, not a preventive-medicine government. Nothing is done until the patient is near death, when everyone panics. Our government reflects our behavior as individuals. How many of us go to a doctor when we are coping with our ailments, when death does not seem imminent? Doctors' services are expensive, as are the needed changes in our use of energy.

# 6

# Global Warming: The Climate Is Changing

We have a fundamental problem. We haven't come up with political institutions that take the long view of serious problems. The car is a major cause of the [federal] deficit and of global warming and air pollution. Any serious attempt to deal with these problems will be painful. Trouble is, we just don't have the political mechanisms to impose pain on citizens in a democratic society.
—Michael Walsh, *Greenpeace*

The data are in. Global warming is a fact, not an unproven scientific theory (figure 6.1). The 24 warmest years since 1900 have occurred since 1973. The 10 warmest years in the instrumental record all occurred since 1990. The year 1997 was the warmest ever recorded until then. This record was beaten in 1998 (the hottest year on record), 2001, 2002 and 2003. The climate of the 1990s was the warmest since at least the Middle Ages. Most of the world's glaciers are melting. On the Antarctic Peninsula the average temperature has risen about 5°F since 1945. In the Arctic the ice cap is melting at a rate that may allow routine commercial shipping through the far north within 10 years. On land, the area of permanently frozen ground (permafrost) has decreased by nearly 30 percent since 1900. The ranges of migratory plants and animals have shifted an average of 3.7 miles per decade toward the poles.[1] In Canada's Yukon Territory, female squirrels are mating 18 days earlier than they were a decade ago because spruce trees are producing more cones, the squirrel's main food, during lengthened summers. The squirrels have more energy to reproduce, according to a squirrel (ly?) biologist. Robins are building nests 250 miles north of the Arctic Circle.

The winter of 1999–2000 was the warmest winter the United States has ever had. Every state in the continental United States was warmer than its

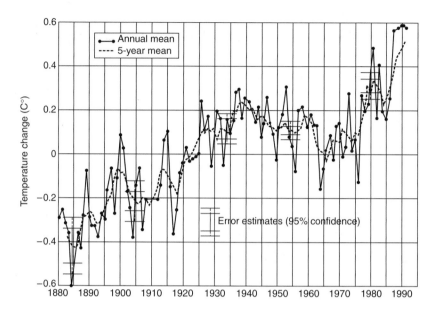

**Figure 6.1**
Average global surface temperature since 1880, a combination of both land and ocean readings (NASA).

long-term average. When "proxy" data from tree rings, pollen, and other sources are considered, 1999 was the hottest year of the millennium. Researchers have calculated that all the glaciers in Glacier National Park will be gone in the next 70 years. The Greenland ice cap is losing thickness at up to 3 feet a year. There is no doubt that the climate is getting warmer and may continue to do so for the forseeable future. Polls indicate that most Americans regard global warming as a serious threat.

The amount of warming of the atmosphere is equivalent to placing two miniature Christmas tree bulbs over every 10 square feet of earth's surface and keeping them burning night and day.

**What Determines the Earth's Surface Temperature?**

We are embarked on the most colossal ecological experiment of all time: doubling the concentration in the atmosphere of an entire planet of one of its most important gases; and we really have little idea of what might happen.
—Paul A. Colinvaux, *Why Big Fierce Animals Are Rare*

The temperature of our air is controlled mostly by the output of energy from the sun. This energy exists as a spectrum of wavelengths whose distribution is determined by the sun's temperature. The hotter the object doing the radiating, the shorter the wavelengths that are radiated. Because the sun is very hot about 90 percent of the wavelengths are relatively "short," with wavelengths less than 6/100,000 of an inch (1.5 micrometers). Except for the few percent of the wavelengths that are less than 1.5/100,000 of an inch (0.4 micrometers; ultraviolet wavelengths), these short waves pass through earth's atmosphere without interference. The waves hit the earth's surface, are absorbed, and are then reradiated back to the atmosphere. But because the earth's surface temperature is much less than that of the sun, the reradiated wavelengths average 15–20 times longer than the ones it received.

So what difference does that make? The difference is that the long wavelengths given off by the earth's surface are massively absorbed by some of the gases in the atmosphere. These gases absorb the earth's heat radiation and thus warm the surface, just as a blanket traps body heat. This is fortunate for us because if the long waves passed uninterrupted through the atmosphere, our surface temperature would be about 0°F rather than the 58°F it is now. This trapping of long wavelengths is what is called the *greenhouse effect.*

### Which Gases Trap the Heat?
Which atmospheric gases are the absorbers? Here we get to the nub of the global-warming problem. Nearly all the absorption of the long wavelengths is accomplished by five gases: water vapor, carbon dioxide, methane, chlorofluorocarbons (CFCs), and nitrous oxide (figure 6.2). The most abundant atmospheric gases, nitrogen and oxygen, have no heat-trapping ability. Water vapor is by far the major absorber, trapping heat about a hundred times more effectively than carbon dioxide. About 60 percent of total absorption is due to water vapor in the air. But its concentration in the atmosphere is not directly affected by human activities. The amounts of the other four heat absorbers, however, are strongly affected by human activities, and all of them have greatly increased in abundance during the past 150 years. We have contributed to, or perhaps caused, an *enhanced* greenhouse effect. Hence, global warming.

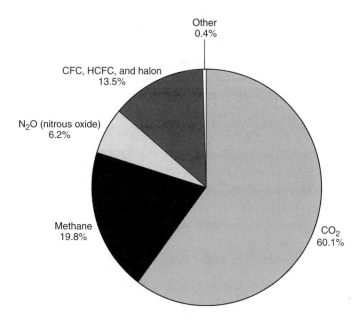

**Figure 6.2**
Estimated contribution of long-lived greenhouse gases to the enhanced greenhouse effect (ICPP, UN, and World Meteorological Organization).

Since 1850 the amount of carbon dioxide emitted into the air by the combination of coal and oil has quadrupled and the amount of carbon dioxide in the air has increased by 31 percent. Assuming that human activities are the sole cause of global warming, the increased amount of carbon dioxide is now responsible for 60 percent of enhanced global warming. And carbon dioxide emissions are on a steep upward trend, largely because of huge increases by developing countries (table 6.1).[2] In the United States, the annual growth rate of carbon dioxide emissions was 1.8 percent between 1990 and 2000, and it continues to rise.[3]

Methane abundance has more than doubled and now causes 20 percent of enhanced global warming. CFCs are a creation of industrial chemists and did not exist until 1928. The abundance of nitrous oxide has increased by 13 percent and is responsible for about 6 percent of our higher temperature. What could have happened since 1850 to generate all this gas? Could it be the increased number of politicians? Tempting hypothesis, but probably not. The answer is clear. America was transformed from a rural and

Table 6.1
Carbon dioxide emissions, 1950–2000

| Country | Billions of tons | Percent |
|---|---|---|
| United States | 181.6 | 32.0 |
| European Union | 127.8 | 22.5 |
| Russia | 68.4 | 12.1 |
| China | 57.6 | 10.2 |
| Japan | 31.2 | 5.5 |
| Ukraine | 21.7 | 3.8 |
| India | 15.5 | 2.7 |
| Canada | 14.9 | 2.6 |
| Poland | 14.4 | 2.5 |
| Kazakhstan | 10.1 | 1.8 |
| South Africa | 8.5 | 1.5 |
| Mexico | 7.8 | 1.4 |
| Australia | 7.6 | 1.3 |
| TOTAL | 567.1 | 99.9 |

*Source: Time,* April 23, 2001, pp. 52–53.

agricultural society to one that is industrial and urban, and from one that depended on human labor for an energy supply to one that depends on coal and oil.

### The Effect of Sulfate Aerosols

Explosive volcanic eruptions commonly contain abundant sulfur and, over billions of years, a permanent layer of sulfate aerosols has developed about 15 miles above the earth. These particles affect the earth's heat balance by scattering solar radiation back to space.

Humans have also contributed to the sulfate in the air by burning sulfurous coal and oil. These additions have had a measurable effect on reducing the enhanced greenhouse effect produced by the greenhouse gases, although the effect of the sulfate aerosols is too small to completely counteract the effect of all the greenhouse gases we have pumped into the atmosphere.

As we will see in chapter 7, sulfate aerosols are the major cause of acid rain and a major contributor to lung disease. Pollution controls are reducing the amount of sulfur emitted from coal- and oil-burning industries and cars.

As sulfur pollution (and pollution by other particulates) declines, the rate of global warming will increase.

The United States is the biggest user of coal and oil (26 percent of world coal use; 26 percent of world oil use) and hence is the largest emitter of heat-trapping carbon dioxide gas and a major producer of soot. Although we use our fuel sources more efficiently than most other industrialized nations, we produce 21 percent of the world's carbon dioxide emissions. Why? Part of the answer is that we are the world's largest industrial economy. Our usage of electricity illustrates this. Electricity is the most important and versatile type of energy in a modern economy, and we are the largest producer of electricity. Although few new coal-fired power plants are being built anywhere in the world, they used to be the only type available. Hence, 50 percent of America's electricity is still generated by the burning of coal. As a result, electricity generation accounts for 35 percent of our emissions of carbon dioxide, 29 percent of the nitrous oxide, and 66 percent of the sulfur dioxide.[4]

## Automobiles

One hundred years ago the 8,000 cars on American roads were considered to be one of the greatest benefits to the environment yet devised. They held the promise of virtually eliminating the filth associated with the mountains of horse excrement clogging the streets from the 4 million horses used for transportation.[5] A century later we are not so sure about the environmental benefits of our most prized possession. We now recognize that gasoline- and diesel-powered cars are essentially global-warming factories on wheels. The average American spends about 440 hours per year behind the wheel of a car, the equivalent of about 55 workdays. Approximately 91 percent of all the person-miles traveled in the United States are in privately-owned automobiles. Car ownership accounts for 15–22 percent of all household expenditures.[6] We burn 65 percent of our oil consumption for transportation.[7] Our cherished automobiles, sports utility vehicles, and trucks depend on gasoline, a refined petroleum product. Our cars and SUVs alone account for 43 percent of the oil we consume. World motor-vehicle production has risen almost every year since 1950 and is now more than 57 million per year. Worldwide, there are 670 million cars on the road, a number predicted by the UN to rise to between one billion and one and a half billion by 2030.[8] Auto sales in China, the world's most populous country, grew 73 percent in 2003[8] compared to about 3 percent in the United States and Europe.[9]

As you might guess, Americans own an inordinate number of the world's cars, about 25 percent of the total. Since 1970 the population has increased by one-third, but the number of vehicles on the road has nearly doubled. In 2000, 18 percent of American households owned three *or more* vehicles;[10] 40 percent own two, 32 percent own only one, and 10 percent of households manage life without a car. According to the Transportation Department, there are now more cars in American households than drivers. We also use our cars the most (2.3 trillion miles in 2000), and hence use an incredible amount of gasoline, about 350 million gallons per day. The typical automobile engine has an efficiency of only 20 percent, wasting 80 percent of its gasoline as heat.

As with coal, the exhaust fumes from gasoline combustion are loaded with carbon dioxide. Motor vehicles in the United States generate about 5 percent of the world's carbon dioxide emissions, more than all of the sources in heavily industrialized Japan. Motor-vehicle exhausts also produce abundant nitrous oxide and, as a noxious bonus, hydrocarbon emissions from cars interact with sunlight to produce ground-level ozone, also known as smog. Motor vehicles account for 28 percent of America's emissions of greenhouse gases.

Motor vehicles contribute another greenhouse (and noxious) material to the air, soot. Recent research suggests that 21 percent of the world's soot emissions, essentially pure black carbon, is emitted in the United States.[11] Soot may be responsible for 15–30 percent of global warming[12] and thus may be second only to carbon dioxide as a cause of enhanced global warming. Soot particles, like carbon dioxide, absorb heat radiated from the earth that would otherwise escape into space. But soot also soaks up radiation from the sun, converting the energy to heat, much as black clothing does on a summer day. Automobile engines, coal-fired generators, and forest fires contribute to the soot problem. Eliminating soot from fossil fuels could reduce global warming by 40 percent within a few years because it stays in the air only a few weeks, unlike greenhouse gases, which remain for many years. Soot removal would also have a beneficial effect on your lungs (see chapter 7).

The scarcity of public transportation in the United States and the well-known American love affair with the automobile guarantee that the familiar scene of gas guzzlers transporting one passenger will not end soon. Eighty-seven percent of us commute to work by car, and three-quarters of us commute alone.

Despite the perception of Americans that gas prices were at an all-time high early in the twenty-first century, we remain uninterested in small cars despite the beneficial effect they have on our bank accounts and on slowing global warming. Sales figures reveal that only 5–6 percent of American car buyers choose small cars, compared to 46 percent who want the larger and less fuel efficient but jazzier SUVs, minivans, and light trucks. Between 1985 and 2000 the market share of small cars decreased from 33 percent to 24 percent. Sales of intermediate and standard-size cars decreased from 31 percent to 17 percent. Cars that weigh 2,500 pounds or less were 18 percent of all passenger vehicles sold in 1985 but are now less than 0.5 percent of sales, according to the EPA. The popularity of gas-guzzling vehicles has reduced the average fuel economy of 2003 model cars to 20.8 miles per gallon, about 6 percent below the modern high point of 22.1 mpg set in 1988. (Henry Ford's 1908 Model T car got 25 mpg.) The continuing increase in popularity of SUVs (25 percent of U.S. light-vehicle sales in 2002),[13] which average only 14 mpg, continues to depress the fuel economy of American vehicles. According to the Department of Transportation, demand for gasoline has increased by 13 percent since 1990. Surprisingly, a 2002 Roper poll revealed that 62 percent of Americans believe the mpg of vehicles in the United States had increased in the previous 10 years. Only 17 percent realized it had gone down.

The Sierra Club has calculated that driving an SUV for 1 year is equivalent to the energy wasted by leaving your refrigerator door open for 6 years, keeping a light in your house on for 30 years, and leaving your TV on for 28 years.[14] Nevertheless, SUV sales more than tripled between 1990 and 2001. The big three automakers were overjoyed at this trend; SUV sales account for about 90 percent of their profits.[15] The current generation of Americans love their SUVs. Perhaps generations now departed would have, too. An article in the July 17, 2000, *New York Times* asked

"Was Freud a Minivan or S.U.V. Kind of Guy?" My personal guess is that he would have gone for a red Porsche convertible.

Because of a combination of federal mandates and the environmental concerns in California and a few other states, manufacturers have been forced to produce hybrid gasoline/electric cars. In the present early stage of development the cars are small and any batteries need frequent recharging. These factors will limit their sales. Much of the electricity used for recharging comes, of course, from the highly polluting coal-burning power plants. Nevertheless, the use of the hybrid cars does result in a net decrease in carbon dioxide emissions. The auto industry views cars powered only by batteries as a dead end and favors hybrid vehicles as the best interim solution until hydrogen-powered vehicles can be mass produced, preferably with refined petroleum products as the hydrogen source.

The best solution to decrease auto emissions would be a massive federal, state, and local investment in public transportation, particularly railroads. But the rail network in the United States decreased by 28 percent between the mid-1980s and mid-1990s.[16] A century ago the United States led the world in public transit. In 1910 almost 50 times more Americans commuted to work by rail than by car, and a decade later almost every major U.S. city had a rail system. But following World War II, the Federal government emphasized construction of roads and freeways, which spelled the death of the public transportation system.

Nevertheless, there are signs that Americans are beginning to realize the benefits of public transport. The number of trips taken via public transport increased by 21 percent between 1995 and 2000, while growth in driving rate increased by 11 percent. In 2000 the number of miles driven by car stayed flat for the first time in 20 years. Transit use grew by 3.5 percent. Subways and light rail have seen the fastest rates of growth, largely due to new systems in Salt Lake City, Denver, and Dallas, and extensions of existing service in Los Angeles and other metropolitan areas around the country.

A team of chemists in China has made a technological breakthrough of great potential importance to our carbon dioxide–polluted age. Not only have they found a way to dispose of our excess gas but they are on the way to creating a money-saving bonanza for couples planning to marry. How, you say? The scientists have succeeded in making transparent and colorless diamonds from carbon dioxide gas (diamonds are composed entirely of carbon).[17] They have repeated their procedure more than 80 times with 100 percent success and no safety problems. Unfortunately, the largest diamond made so far weighs only .02 carat. But research is continuing. Talk about your win-win situations! Clean air and engagement rings too.

### Underground Coal-Mine Fires

A source of greenhouse gases rarely mentioned in considerations of global warming is fires belowground in unmined coal beds and abandoned coal mines. These fires produce nearly as much carbon dioxide as is emitted each year by all the cars and small trucks in the United States.[18] There is geological evidence that lightning and spontaneous combustion of coal have spawned such fires for hundreds of thousands of years. Spontaneous combustion occurs when the mineral pyrite (iron sulfide), common in coal beds, and other reactive minerals in coal are exposed to oxygen in the air. They begin to release heat, which, if not dissipated by air currents, builds until the coal ignites. One fire in Australia is believed to have been going strong for 2,000 years. Mine fires are difficult and expensive to stop. Most are simply left to burn themselves out when the coal bed is finally consumed, a process that may take many hundreds or thousands of years.

### Methane

Methane, the second most important greenhouse gas, is produced when organic matter decays in the absence of oxygen. The largest contributor of methane is, believe it or not, gaseous emissions from the both ends of the earth's 1.3 billion cattle. This is closely followed by rice paddies and by combustion of oil, natural gas, and biomass (figure 6.3). Together, these sources account for more than two-thirds of anthropogenic methane production.

Most of these sources are difficult or impossible to control. Cattle burping (90 percent of methane emissions) and flatulence (10 percent) is largely uncontrollable by either the cows or us, although a vaccine developed recently in Australia appears to reduce ruminant flatulence by 20 percent.[19] The government of New Zealand in 2003 proposed a "flatulence tax" on methane emissions from the country's 10 million cows and 45 million

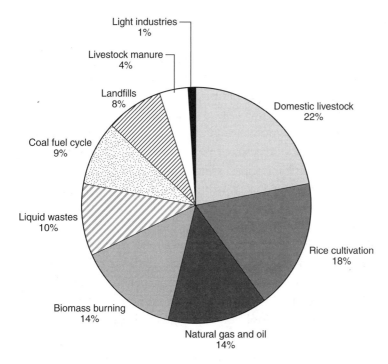

**Figure 6.3**
Sources of anthropogenic methane (Environmental Protection Agency).

sheep, the source of 40 percent of the nation's methane emissions. The tax would be 49 cents on each cow and 6 cents (U.S.) on each sheep. In protest, farmers organized a "Raise a Stink" campaign and sent parcels of sheep and cow manure to their country's lawmakers. Postal workers complained that their health was being endangered. Massive street protests and a petition from half the nation's farmers caused the government to withdraw its proposal. In the United States, farms are largely exempt from the provisions of the Clean Air Act and have no incentive to reduce farm air emissions. We can safely assume they would not favor a flatulence tax.

Rice cultivation and associated methane emissions are increasing to feed the exploding population in China and other parts of eastern Asia. As noted earlier, coal and oil need to be replaced as major energy sources, but this will not happen for many decades (see below). One important source of methane not considered in the EPA analysis is natural wetlands, stagnant areas (like rice paddies) essential to coastal ecology and flood prevention and now under federal protection.

Some of the biomass burning in the United States consists of forest fires, now thought to be essential for the long-term ecological health of the forest and, so long as they do not endanger people or their property, should be allowed to burn themselves out. The fires set deliberately in tropical areas can in principle be halted, but in practice it is extremely difficult because of political and social problems in impoverished tropical countries. Animal wastes are increasing rapidly in abundance as meat consumption increases in concert with the increasing standard of living in much of the world. For example, meat consumption in China, which has 20 percent of the world's population, increased by 50 percent between 1990 and 1995 as the Chinese public slowly reaps the financial, cholesterol, and saturated-fat benefits of their rapidly growing economy.

Emissions of methane are expected to increase by 45 percent in the next 50 years. This is troubling not only because of the numerical increase but also because each methane molecule is 25 percent better at trapping the long wavelength emissions from earth's surface than is each molecule of carbon dioxide. On the other hand, the lifetime of a methane molecule in the atmosphere is only 12 years, compared to 60 years for a molecule of carbon dioxide. And the amount of carbon dioxide in the air is increasing rapidly. As a result, methane will be responsible for less of the global-warming problem in 2050 than it is today. Carbon dioxide is the main problem today and will continue to be in the future as well.

## CFCs

Chlorofluorocarbon gases (CFCs) were invented in 1928 and have found many uses. Because of their unusual chemical and physical properties, they are used as coolants in refrigerators and air conditioners (Freon), as cleaning agents for electronic components, as aerosol propellants in spray cans, and in styrofoam drinking cups and insulation. In 1987 it was determined that CFCs attacked and destroyed ozone in the stratosphere, and for this reason production was banned shortly thereafter. The role of CFCs as a contributor to global warming was not a factor in the decision to ban them but, hey, we'll take help from wherever it comes. Now that the production of CFCs has almost ended, they will disappear from the atmosphere in about 50 years. So at least one contributor to the global-warming problem is on the way to being eliminated.

## Nitrous Oxide

Nitrous oxide is a greenhouse gas that, like the others, has been increasing for the past hundred years. Its abundance in our atmosphere is tiny, however, only 3/100,000 of 1 percent. Hence, even though it traps the earth's heat emissions a few hundred times more effectively than carbon dioxide, it is still only fifth in importance as the cause of the enhanced greenhouse effect. Most of the increase in nitrous oxide during the past century is attributed to increased use of nitrogen-containing fertilizers.[20]

The environmental benefit of using natural gas rather than coal to produce electricity in power plants has not been lost on the utility industry. New power plants in the United States are increasingly gas-fired. In older power plants, the installation of modular gas turbines gives power companies the capability to add more power-generating capacity in small increments to more closely match growth in demand.

## Urban Sprawl

Suburbia, the middle ground between nature's beauty and civilization's conveniences, has been viewed as the promised land by millions of Americans over the last several generations.

—Philip Dolce, *Suburbia*

Since World War II Americans have deserted the cities in favor of the less congested suburbs. In the largest metropolitan areas suburban population has grown 10 times faster than the central-city population. Only a third of this can be traced to population growth. More of our population now lives in the 'burbs than in the metropolitan areas. Phoenix, for example, is developing land at a rate of 1.2 acres *per hour.*[21] In America hardly anyone says no to a suburban developer. Sprawl is, in the long run, probably irreversible. If you can afford it, and an increasing proportion of Americans apparently can, owning your own home on a large plot of ground is emotionally more satisfying than living in an apartment in a large building in the city. Despite newsworthy occasional and temporary reversals in some cities, urban flight has not stopped. And the public is not likely to vote for zoning laws that mandate multifamily housing in suburban areas.

Among the many negative effects suburban sprawl has on the environment (gobbling up farmland, forests, floodplains, wetlands, and green space) is the effect on pollution. Because of inadequate public mass transportation, sprawl lengthens car trips and forces us to drive everywhere. Residents of sprawling communities own more cars and drive three to four times as much as those living in compact, well-planned areas.[22] Studies show that adding new lanes and building new roads only lead to more traffic, which, of course, means more fuel use and more air pollution.

Another problem with urban sprawl is that individual suburban homes are typically much larger than apartments in city dwellings and hence require twice the energy for heating and cooling, energy generated using fossil fuels. Urban sprawl is a major contributor to America's air-pollution problem.

### Global Warming and Precipitation

The belief that we can manage the earth and improve on Nature is probably the ultimate expression of human conceit, but it has deep roots in the past and is almost universal.
—Rene Dubos, *The Wooing of the Earth*

Heat causes water to evaporate, and the hotter it gets the more water will evaporate. This will magnify the enhanced greenhouse effect (positive feedback because water vapor is the major absorber of the earth's infrared radiation.) Hence, we expect global warming to increase the amount of evaporation from the oceans and consequently to increase world rainfall (and humidity) but to decrease snowfall. These changes have already been detected. Since the 1960s the earth's snow cover has decreased by about 10 percent and the thickness of the Arctic ice cap has decreased by an astonishing 42 percent. Water discharge from the rivers draining northward from Eurasia into the Arctic Ocean increased by 7 percent between 1936 and 1999.[23] Data from more than 5,000 rain gauges on six continents indicate that average annual rainfall has increased by almost an inch during the past hundred years.[24] However, the increase has not been uniformly distributed. Rainfall has decreased in the tropics but increased at high latitudes.

Over North America the increase has been more than an inch and a half. Although it is clear that rainfall has increased in the United States (figure 6.4), the increase has not been evenly distributed. The northern Rockies, Texas, and California have been drying out while the vast midcontinental grain belt has benefited. An inch or so more rainfall is certainly beneficial to our main agricultural breadbasket, but when it takes a hundred years to materialize, the effect on crop production is hardly noticeable. And the decreased precipitation in the California fruit and vegetable belt further reduces the net impact on crops. At present, there is no explanation for the uneven distribution of the increase in rainfall over the continental United States.

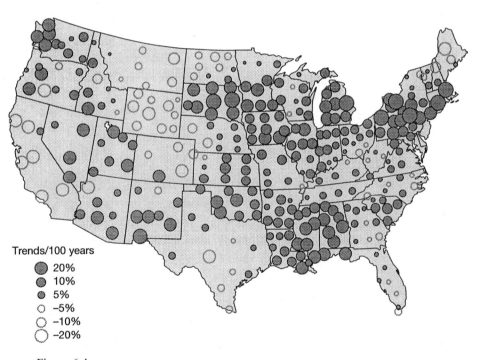

Figure 6.4
Trend in annual rainfall in conterminous United States since 1900 (T. R. Karl, R. W. Knight, D. R. Easterling, and R. G. Quayle, "Trends in U.S. Climate during the Twentieth Century," *Consequences* 1, no. 1 (1995): 5).

One result of the increased precipitation and humidity that is commonly forgotten in discussions of the enhanced greenhouse effect is the fact that water vapor is a potent greenhouse gas. It is the major cause of the normal greenhouse effect that keeps the earth's surface temperature in the desirable range. If there is more moisture in the air the currently enhanced greenhouse effect will be further enhanced, although there is no quantitative evaluation available yet. The warmer it gets, the more moisture will be in the air. The more moisture there is in the air, the warmer it gets.

## Global Warming and the Oceans

Climate is an angry beast, and we are poking it with sticks.
—Wallace Broecker, oceanographer and climate scientist

The effect of global warming on the oceans can be considered in two interconnected parts, the effect on sea level and the effect on ocean currents. The effect on sea level has received considerable attention from climate scientists. Possible changes in ocean circulation have received less attention, not because they are less significant, but because the basic controls of oceanic circulation are still too poorly understood for possible modifications to be evaluated.

### Sea-Level Rise
The sea level is rising, mostly because of the melting of Antarctic and other glacial ice that started 18,000 years ago. A secondary cause is the expansion of ocean water due to its increasing temperature.[25] Continued temperature increase will accelerate these trends, putting more water in the ocean, causing further sea-level rise. The effect is already measurable. Since 1750 sea-levels along the coasts of Nova Scotia and Maine have risen 12–24 inches.[26] During the past hundred years mean sea level in the world ocean has risen about 7.5 inches, not catastrophic for most of us. But several investigators have predicted an additional sea-level rise of 4–34 inches by 2100. Such increases would be disastrous for people who live in coastal cities. Because of the long lifetimes of greenhouse gases, if greenhouse emissions were stopped today, sea level would continue to rise for at least 200 years, though at lower rates.[27]

The rising sea level is on a collision course with the stampede of Americans to the salty air. Of the 20 fastest-growing counties, 17 are coastal. In year 2002 dollars, the federal government has spent $3.7 billion since 1923 on beach replenishment.[28] Property values at risk today along the Atlantic and Gulf coasts exceed $3 trillion and continue to rise as coastal development proceeds.[29] Beachfront homes afford scenic views, but wave and current action associated with the tide causes extensive damage. Water weighs 1,700 pounds per cubic yard and extended pounding by frequent waves can demolish any structure not specifically designed to withstand such forces. The Federal Emergency Management Agency (FEMA) estimates that one-fourth of the buildings within 500 feet of America's coastlines are threatened by sea-level rise in the next 60 years. Nearly 87,000 homes and other buildings are likely to wash away.[30] Louisiana is already losing 1 acre per 24 minutes, 34 square miles every year to a combination of subsidence and sea-level rise.[31] At that rate the entire state and Cajun food will be a memory in 1,200 years. There already are more than two dozen flood-control structures designed to protect New Orleans against hurricane waters and Mississippi River floods. New Orleans is the nation's fourth busiest port and already averages 8 feet below sea level.

Erosion by a landward-moving shoreline is slow but relentless. Its impact is illustrated by the problem of the Cape Hatteras Lighthouse, which was 1,500 feet from shore when erected in 1870. By 1987 the lighthouse was in danger of falling victim to erosion by the advancing ocean. Finally, in 1999, the National Park Service moved it inland 2,900 feet at a cost of $9.8 million. Few people can afford to move their homes.

A verbally violent debate has been going on for decades about what to do about shoreline (beach) erosion. Should America's beaches be protected, at least temporarily, by structures such as seawalls, groins, and jetties? Should sand from offshore or from far inland be transported at great expense to save the beach? Or should we simply face the inevitable and let sea-level rise take its natural course? Since the 1980s, the U.S. Corps of Engineers, the government agency responsible for protecting our beaches, has put replenishment before seawalls and other structures as the preferred response to erosion. But replenishment has proven to be costly, temporary, and unpredictable because shoreline dynamics, the interaction between land and sea, are poorly understood. Mother Nature will not be defeated. Ultimately, humans will have to accept the fact that beaches change position. They don't disappear. They simply move landward. It is inevitable and the sooner we learn to adapt by limiting construction near shorelines and moving existing structures where possible, the better.

Coastal dwellers are a large part of the world's population; many of the largest cities are located on coastlines. Between 15 and 20 percent of low-lying Bangladesh will be under water in 50 years and perhaps one-third of the country will disappear by 2100. It is certain that several small Pacific island nations will disappear. Because the lifetime of carbon dioxide molecules in the atmosphere is 60 years (it is removed by photosynthesis), it is already too late to save these nations, even if proposed decreases in carbon dioxide emissions materialize.

Should the earth's large ice sheets melt completely, sea level would rise about 150 feet. About 15 percent of the continental United States would disappear (figure 6.5). The states of Massachusetts, Connecticut, Rhode Island, New Jersey, Delaware, Florida, and Louisiana would be completely under water, as would our nation's capital. Say goodbye to New York City, Philadelphia, Miami, New Orleans, Houston, San Francisco, and Seattle. Albany, Raleigh, Little Rock, and Sacramento will be seaports. The mind boggles.

The focus of scientists on ice melting tends to focus on the West Antarctic ice sheet, which has a volume of 800,000 cubic miles. It is important not only because of its size but also because it sits on ocean-floor rock rather than floating. Marine ice sheets persist only as long as they are large enough to squeeze out underlying water, which makes them inherently unstable. Should this ice sheet float free, as other marine ice sheets have done, global sea level would rise 15–20 feet, innundating most of Florida and hundreds of low-lying cities in the United States and elsewhere. Since 1990, 36 cubic miles of ice have melted from glaciers in West Antarctica. Climate scientists differ on when, or if, the Antarctic ice will melt completely. But few of them entirely discount the possibility. In 2000 an ice sheet 185 miles long and 23 miles wide, just slightly smaller than Connecticut, broke off from Antarctica to form an iceberg. In 2002 an iceberg more than twice the size of Manhattan broke loose, as did another twice the size of Luxembourg. Scary stuff. As one coastal-management expert

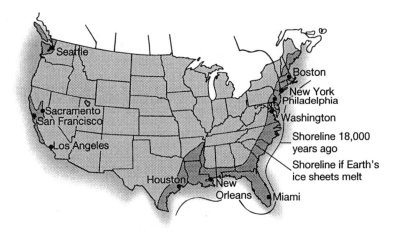

**Figure 6.5**
Location of the shoreline 18,000 years ago at the height of the last glaciation in relation to the present shoreline. Also shown is the shoreline if all existing ice were to melt. Many states would disappear.

has said, "Most government agencies are doing nothing more than rearranging the deck chairs on the Titanic."

**Ocean Circulation**
The major ocean currents are driven by three things—the rotation of the earth, the temperature difference between the poles and the equator, and salinity differences between one part of the ocean and another. Global warming will decrease the temperature difference between the equator and the poles. Warming of the oceans will cause Antarctic ice to melt and put a large amount of freshwater into the world ocean. The effect such changes will have on oceanic circulation is uncertain, but climate scientists believe there is cause for concern.

A key concern is that major heat-transporting currents such as the Gulf Stream may weaken, with serious to catastrophic consequences for the Eastern United States and Europe (figure 6.6). According to most climatologists, the Gulf Stream and its extension, the North Atlantic Drift, supply Western Europe with free heat equivalent to the output of about a million power stations.[32] Without the Gulf Stream the average temperature in Europe might drop by as much as 17°F. After all, Ireland and the United King-

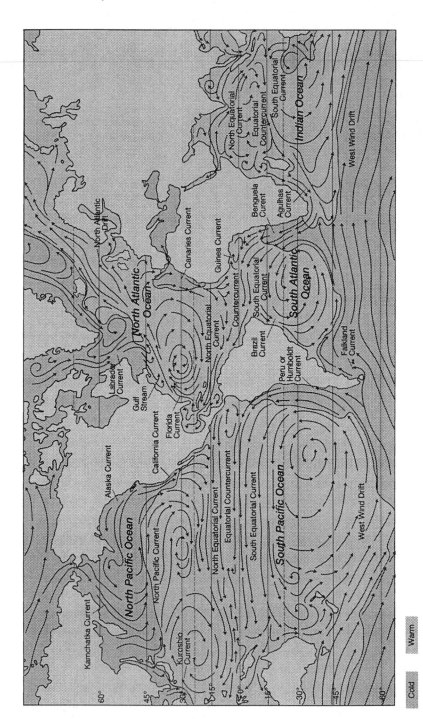

**Figure 6.6**
Average position and extent of principal surface currents in the world ocean.

Cold    Warm

dom are at the same latitude as Labrador in Canada. Ireland and the United Kingdom are less than half the size of Labrador but have 2,000 times the population. Current climatological theory holds that the equatorial heat carried north by the Gulf Stream keeps Britain habitable (also foggy and rainy). Perhaps even more frightening is the possibility, suggested by studies of the changing composition of gases in air bubbles in Greenland ice in relation to the age of the ice, that the change might happen within 10 years or less, leaving no time for people to adjust.[33] The National Academy of Sciences says that the rapid change occurs in the same way as the slowly increasing pressure of a finger eventually flips a switch and turns on a light. Once the switch has occurred, the new hostile climate lasts at least for decades, and possibly centuries. At present, no one knows whether such a catastrophe will happen, but many climate modelers are concerned. It may be significant that Scandanavia is the only place in the world where temperate mountain glaciers are advancing rather than retreating, a possible indicator of lower temperatures beginning to take hold.[34]

The importance of the Gulf Stream–North Atlantic Drift circulation to England has recently been challenged, however.[35] A computer simulation of atmospheric and oceanic circulation in the Northern Hemisphere indicates that the famous ocean current has no effect on the climate in Ireland and the United Kingdom and that other factors are more important.

Some ocean currents may already be shifting position. In late 2000 hundreds of penguins washed ashore near Rio de Janeiro, 2,000 miles north of their usual haunts. Some climate scientists have suggested that the wayward birds may be a sign that global warming has nudged the cold water currents that the penguins follow for food thousands of miles off course.[36] This should cause some sleepless nights in Ireland and the United Kingdom.

## Global Warming and Human Health

It is clear that small changes in the past 10,000 years have had very large ecological effects and they can happen bloody fast.
—Reid A. Bryson, *Wall Street Journal*

Within the past few years there have been a spate of scary headlines suggesting that a warmer climate may lead to epidemics of malaria, cholera, yellow fever, and other maladies. Typical headline: "Report says global warming poses threat to public health." Another says: "Should we fear a global plague? Yes—disease is the deadliest threat of rising temperatures." Are such worries justified?

For most of the world the answer is yes. For the United States, probably not. According to a medical epidemiologist at the Centers for Disease Control, cholera has been introduced into the United States several times in the past few years; it did not spread, simply "because we have a public health and sanitation infrastructure that prevents it." Lifestyle and public health measures such as mosquito control far outweigh any effects of climate. Another CDC epidemiologist points out that the mosquito vectors of malaria, dengue (an infectious tropical disease transmitted by mosquitoes), and yellow fever have been in the United States for centuries, but the epidemics they once caused have vanished due to mosquito control, eradication programs, piped water systems, and changing lifestyle. They point out that the 1995 dengue epidemic that rolled through Mexico died at the Rio Grande. There were more than 2,000 confirmed cases in Reynosa, Mexico, but only 60 cases across the river in Texas. In summary, we have many more pressing health concerns in the United States than the unlikely possibility of tropical epidemics decimating the population as a consequence of global warming.

Excessive heat does, however, hasten death among the elderly and others whose health has been compromised for medical reasons. The heat wave in France in August 2003 was blamed for 14,800 deaths (estimated by the nation's largest undertaker, who should know). In Italy the count of excess deaths exceeded 4,000. Excess European deaths totaled 35,000. In the United States, heat waves kill more people than hurricanes, tornadoes, earthquakes, and floods combined. And just as extreme precipitation events increase in number with global warming (chapter 2), so do extremes of heat.

One "epidemic" that may not be preventable is an increase in the percentage of male babies over female.[37] A medical researcher has charted births in Germany between 1946 and 1995 and calculated the average temperature at their probable times of conception. His data indicate that warm temperatures favor the conception of boys over girls. Apparently the Y chromosome, which determines the male sex, is better able to resist heat than the X chromosome, which determines the female sex. The investigator concluded that global warming could change the birth ratio by a couple of percentage points in favor of boys.

## Global Warming and Wine

It's a naive domestic Burgundy without any breeding, but I think you'll be amused by its presumption.
—James Thurber, cartoon caption in *The Thurber Carnival*

The summers from 1994 to 2003 were the hottest in Europe in more than 500 years, and, for wine afficionados, this has ushered in a bonanza of exceptional vintages.[38] As a rule, hot summers and early harvests produce great wines, and the 2003 summer produced the earliest harvest since 1893. As noted by a vintner in Germany, "Just look at the row of fine vintages we've had. From 1988 through this year [2003] it has been strikingly warmer than any time I can remember. Everybody talks about it here." Piedmont in northwest Italy had a great vintage every year from 1995 to 2001. In Oregon, the run of exceptional vintages began in 1998. According to one vintner, "In Oregon, the saying used to be you got two really good vintages in 10 years, and in the last 10 years we've probably had nine." In Champagne, France, where single-vintage bottlings were once the exception, done only in the best years, vintages were declared nine times in the decade from 1990 to 1999, as against six in the 1980s and four in the 1970s. To your health!

## Predicting Climate Change

Of course, whether to act is not a scientific judgment, but a value-laden political choice that cannot be resolved by scientific methods.
—Stephen Schneider, climate scientist

Humans like constancy and stability in their lives. Unexpected events such as floods are disturbing, as are occasional crop failures, rising gasoline prices, and our declining physical appearance as we age. The climate we are used to is unconsciously accepted as "okay," if not perfect. We know how to deal with it. But the earth's climate has changed significantly during the past few hundred years and is changing rapidly now. What does the future look like?

A look back at figure 6.1 is useful. After a period of irregular temperature increase from 1885 to 1900, the average temperature dropped
0.23°C (about 0.4°F) in only 4 years, between 1900 and 1904. Then, from
1917 to 1925, there was a very irregular increase of half a degree C (nearly
1°F). After a relatively stable period from 1925 to 1975 there began another erratic rise to a current level that is higher than at any time in the
past 140 years. For how long will this increase continue? Will it moderate
or perhaps reverse as it did between 1900 and 1904? When? Climatologists are uncertain.

The uncertainty stems from unexplained changes in climate before we
started pouring carbon dioxide into the air, before the industrial revolution. Between 800 and 1650 the average temperature in Eastern Europe
varied over more than 1°C (about 2°F), from a so-called Climatic Optimum to a Little Ice Age that saw the expansion of glaciers, the extinction
of farming in parts of northern Europe, and the end of the Viking colony
in Greenland. This drop in global temperatures occurred very rapidly, in
only 10–20 years. The reason for the sudden temperature variation of 2°F
almost overnight is unknown. But it is clear that *something* besides human
activities is kicking the thermometer around. One current theory centers
on the interaction between cosmic radiation and variations in the amount
of solar wind (low-energy charged particles that stream from the sun). Another theory relates changes in weather and climate to variations in mag-
netic activity in the outer layers of the sun. A small minority of scientists
believe that variations in solar activity are the chief cause of the current
episode of global warming, and that the correlation between the human
generation of greenhouse gases and global warming is coincidental. As
statisticians commonly say, correlation is not causation.

Different computer models of future changes in climate give somewhat different answers depending on the assumptions made. It is widely
believed that human industrial activities are a major contributor to the
trend of rising temperatures since 1900, and industrial activities using
fossil fuels are assumed to continue expanding indefinitely into the future. Hence, all computer models project a continuation of rising temperatures. The graphs we see in the media are projections of existing
trends, not predictions of future temperatures. Trends based on predictions of human activities necessarily contain a "what if" emissions sce-

nario. For example, we repeatedly read of projected temperatures in 2050 and 2100. But it is almost a certainty that the "business-as-usual" scenario for the next hundred years, the assumption on which the temperature estimates are based, is false. It assumes that renewable and nonpolluting sources of energy will form the same minor percentage of our energy in 2050 and 2100 that they do at present. No scientist or climate modeler believes this, but scary projections sell newspapers. Science cannot provide infallible answers to the questions about future climate that are asked by governmental policymakers. As with the stock market, hindsight is 20–20. Foresight comes with cataracts.

## Global Treaties

All our leaders now call themselves environmentalists. But their brand of environmentalism poses very few challenges to the present system.
—Ivan Illich, *New Perspectives Quarterly*

Within the past 10 years there have been many meetings of world leaders concerned about the possible calamitous consequences of global warming. The principal one was held at Kyoto, Japan, in 1997. The objective has been to reduce world emissions of heat-trapping gases. Some well-intentioned but unenforceable agreements to reduce emissions by specified target dates have been signed, but there is great uncertainty about whether many nations will want to, or be able to, comply with the documents their representatives have signed. The United States and Australia have refused to ratify the treaty they signed at Kyoto.

In desperation, most states have instituted programs to cut greenhouse gas emissions. Twelve states that contain one third of U.S. residents filed lawsuits in 2003 to compel regulation of national carbon dioxide emissions on the basis that current federal policy is jeopardizing the health of the states' citizens. It is possible to imagine similarly motivated lawsuits against the government based on the increased flooding that may be resulting from global warming, or against the automobile manufacturers for pushing their very profitable but highly polluting SUVs. The success of these lawsuits against the federal government or industry is questionable, but I sense a state of financial euphoria growing in the legal profession.

About half of European Union countries are failing to meet their renewables targets, according to the World Wildlife Fund, European Environment Agency, and EREF, the European green-energy industry group.
By 2010 the EU is supposed to reduce greenhouse-gas emissions by 8 percent from 1990 levels and increase their share of renewably produced
power to 22 percent of their total energy consumption. With current policies the 15 European countries will achieve only 15–17 percent. As of June
2003, Germany, the United Kingdom, France, Sweden, Luxembourg,
Denmark, and Ireland are on track to achieve their emissions targets but
the others are not.[39]

Although most countries agree that emissions of carbon dioxide from
industrial and agricultural sources need to be reduced, there is a wide difference of opinion about how to accomplish this. Which nations should
bear the deepest cuts? One basic split is between the developed ("rich")
and "developing" ("poor") nations. The less-developed nations want the
industrialized nations to bear most or all of the reductions. After all, the
developing nations say, "You have produced the present crisis [table 6.1]
so you should solve it. We need industrialization to become as well off as
you are."

The developed nations believe that agreeing to do all the emissions cutting would be disastrous for their economies. We would end up drinking
warm beer in a cold, dark house. The developed nations say "We will cut
some, but you must also cut. You now account for about 40 percent of
current emissions of carbon dioxide and your rapid growth indicates that
within 20 years you will be producing as much greenhouse gas as we do.[40]
We will not agree to suffer unilaterally." In the case of the United States, a
treaty to cut emissions of greenhouse gases, which have risen 12 percent
between 1990 and 2001,[41] must be ratified by Congress and signed by the
president, which will not happen unless the less developed nations agree
to do some cutting as well. Meanwhile, world emissions of carbon dioxide continue to increase, by 1.2 percent between 1999 and 2000, by another 1.0 percent between 2000 and 2001, and still another 1.0 percent
between 2001 and 2002.[42] Apparently it is not only Nero who fiddles
while Rome burns.

The fear of some developed nations, particularly the United States, that
they would end up drinking warm beer in a cold, dark house if emissions

are controlled is doubted by many American economists and refuted by experience in the United Kingdom. Emissions of carbon dioxide fell by 9 percent from 1990 to 2001, despite a 30 percent growth in GDP. According to Energy Minister Brian Wilson, this shows that economic growth does not have to be at the expense of damaging the environment.[43]

The attempt to control global warming by international treaties is in even worse shape than the disagreements between rich and poor nations suggest. The emissions cuts agreed to at international conferences are very small compared to what is needed and would not make a significant dent in the global-warming trend. The United Nations meeting of 175 countries in Kyoto, Japan, in 1997 produced a treaty aimed at reducing greenhouse-gas emissions from industrialized nations by an average of 5.2 percent from 1990 to 2010 (7 percent for the United States). This would reduce by only 0.1° the projected rise of 2.5° by 2050. According to the UN's Panel on Climate Change it would take a 60 percent reduction from 1990 emissions totals to halt global warming. The UN Framework Convention on Climate Change estimated in 2003 that at present emission rates, emissions of greenhouse gases in 2010 will be 17 percent *above* 1990 levels. There is a lot of political posturing going on at meetings concerned with climate change. The Kyoto treaty was intended as a first step to run until 2012. Negotiations for more drastic cuts in emissions after 2012 are due to start in 2005.

## Conclusion

Although global warming may be favored by the 35–40 percent of Americans who live above 40° latitude or at high elevations, and by the 20 percent of the world's population who join them in suffering from winter cold, most people are not pleased by the prospect of continued global warming. They want an end to our enhancement of the greenhouse effect.

There is no doubt that the enhanced greenhouse effect—caused, at least in part, by human activities—is real. Most of the global temperature increase resulting from human activities is caused by the combustion of coal and oil. These fuels have been the foundation of our industrial civilization for 150 years, not only in America but also throughout the world. Changing this will not be easy and cannot be done overnight. But most Americans favor cutting our greenhouse-gas emissions, and the rapid growth in recent years of alternative energy sources elsewhere bodes well for the future. In any event, cutting greenhouse gases will require worldwide cooperation. Air circulates.

The importance of carbon dioxide as the major greenhouse gas will grow in coming decades with the decline of methane as a secondary source and the phaseout of CFCs. The only way to cut emissions of carbon dioxide (and soot) is to switch to other sources of energy for industry, agriculture, and automobiles. Some progress in this direction is being made by the increasing use of natural gas rather than coal in power plants and by the introduction of hybrid and battery-powered vehicles to replace those powered by gasoline alone. Cars powered by fuel cells appear to be the next major step in decreasing greenhouse emissions. Much more needs to be done. But change will be slow, given the inertia of large existing institutions.

A rise in sea level of a foot or more over the next 50 years is very likely no matter how fast the world cuts emissions of carbon dioxide. Perhaps most frightening is the possibility of a large and rapid increase in sea level if Antarctic ice suffers rapid melting. There also may be major disruptions of ocean currents, with uncertain consequences for temperature regimes in the United States and other parts of the world.

The health of Americans is unlikely to be affected by global warming because of our superior sanitation infrastructure. Our water is fairly clean and we can afford whatever disease-eradication measures are necessary to prevent the spread of tropical diseases.

# 7

## Air Pollution and Your Lungs

There's so much pollution in the air now that if it weren't for our lungs there'd be no place to put it all.
—Robert Orben

Long, long ago, on a planet far, far away, the air was pure. It contained only nitrogen (78 percent), oxygen (21 percent), argon (1 percent), and trace amounts of other gases necessary to make complex life possible, such as water vapor (0–6 percent), carbon dioxide (0.034 percent), ozone, nitrous oxide, and a few others. But then the planet's inhabitants, believing themselves of superior intelligence and recognizing that about 10,000 quarts of air and several billion dust particles enter a person's nose and mouth every day, decided to not only increase the amounts of some of the trace gases but also to add other, more creative things. They increased the amounts of carbon dioxide, ozone, sulfur dioxide, nitrogen dioxide, and carbon monoxide. Some gaseous organic compounds were thrown in as well. Then, to spice up the mixture, they added to the air they breathed particulate materials such as microscopic soot (black smoke), lead, asbestos, rubber, arsenic, cadmium, mercury, and other interesting substances.

The results were predictable. There were great increases in respiratory diseases such as bacterial infections, bronchitis, allergies, and asthma. The incidence of eye infections also rose. Babies born in the areas where the air had undergone the most change had smaller heads, lower weight at birth, damaged DNA, and increased rates of birth defects. Many died. Children had smaller lungs. Respiratory disease became the greatest killer of children on the planet. Elderly people also died prematurely, often from heart attacks because their weakened lungs could not process enough oxygen. Bus drivers in urban areas had increased cancer rates, chromosomal abnormalities, and DNA damage from breathing diesel fumes all day.

Crop yields were reduced as pollutant haze reduced the penetration of sunlight and decreased photosynthesis, the process by which plants use sunlight to convert carbon dioxide and water to food and plant fiber. Fruit size and weight decreased. Constituents in the haze affected plant metabolism. Market value was reduced because of spotting on leaves and fruit. Plant death in the field increased as the vegetation became more vulnerable to injury from diseases and pests.

Clearly, experimentation with the atmosphere in which they had evolved was not a good idea for the people of this planet. Because of their superior intelligence they decided to scale down or end the experiment. But it would not be easy.

**Where Is This Place?**

We all live downwind.
—Bumper sticker

As you no doubt have surmised, the long, long ago, far, far away planet is earth in the twenty-first century. Everything described is true. Air pollution has harmed crop production and causes more deaths in the United States than traffic accidents. Traffic fatalities total just over 40,000 per year, while air pollution claims 70,000 lives annually.[1]

Air circulates, so Americans do not live in an atmospheric cocoon. We are not immune to the effects of air pollution arising in other parts of the world. And in many other parts of the world the air is so bad that people walk around wearing surgical masks to filter out some of the particles. Most urban children in the "developing countries" inhale the equivalent of two packs of cigarettes each day just by breathing. "Oxygen bars" have opened in Los Angeles, New York, Mexico City, Beijing, London, and Tokyo. In Los Angeles, for $15 you can breathe lemon- and lime-scented clean air briefly to let your lungs know what they are missing ($13 without the citrus). In Beijing a slug of good air is only $6, presumably because of lower production costs. In Mexico City the cost for a supplement of commercial oxygen is $2 per minute. A gambling casino in Cripple Creek, Colorado, offers oxygen hits in eight delicious flavors, including peppermint and tangerine. And air pollution is getting worse worldwide as impoverished nations such as India and China attempt coal-based rapid industrialization in their increasingly urbanized and car-using societies. The world's worst air is in Dacca, Bangladesh. It replaced the former leader, Mexico City, in 1997.

What are the major pollutants, how did they get into our air, and what can we do about getting them out? What do American political leaders, both federal and local, propose to do to clean our air? Have they made much progress? What can we hope for in the near future? Do we want to encourage enterprising entrepreneurs to increase the number of oxygen bars in our major cities? Should we impose a graduated tax on people who breathe cleaner air to pay for cleaning up the air in dirty cities? It could work like the federal income tax. If the air where you live is unusually clean (as it is for penguins in the Antarctic) your tax bill would be high. If your air is unusually dirty (Houston, Los Angeles, New Orleans) you would be exempt from paying the tax. Clean air is already traded on stock exchanges in Australia and at the Chicago Climate Exchange.

**Pollutant Gases In The Air**

When you can't breathe, nothing else matters.
—American Lung Association

Each of us takes a breath every 4 seconds, about 8.5 million breaths per year. When resting, you and I inhale about 2,500 gallons of air per day, 7 quarts per minute. That's 1,500 trillion (1.5,000,000,000,000) gallons per day for the world's population of 6 billion. The molecules of air we breathe were at one time in the lungs of Beethoven, Napoleon, Attila the Hun, Jesus, and Moses. When doing heavy work—that is, huffing and puffing—the daily amount for each person rises from 2,500 to 15,000 gallons. An average American breathes 3,400 gallons.[2] We use a lot of air. It would be nice if it were clean. Should a "Right to Breathe" be added to our Bill of Rights?

The noxious gases in our air that are monitored by urban air-quality authorities are sulfur dioxide, nitrogen oxide, ozone, and carbon monoxide.

In 2004, more than half of all Americans lived in counties that did not meet the EPA standard for at least one of these pollutants. The problem is worldwide. Twenty of the 24 megacities of the world (more than 10 million people) do not pass muster for all of these gases and lead. A few cities have succeeded in improving their air somewhat during the past few years, but in most cities the air has continued to worsen. In Athens the death rate jumps by 12 percent when the level of sulfur dioxide exceeds a critical threshold.[3] In New Delhi, 20 percent of the traffic police at busy intersections need regular medical attention for lung problems. Soon the number of coughs per hour in these cities may rival the number of words spoken.

Perhaps more distressing than increases in coughs is the effect of traffic fumes on male sperm and, therefore, reproductive success. In a study of male toll-booth attendants on Italian motorways, a recent investigation found the men had poorer-quality sperm than other young and middle-aged workers in the same area.[4]

## Sulfur Dioxide

Two-thirds of the noxious sulfur compounds in the air of North America have been produced by human activities,[5] mostly by the combustion of coal and oil. Coal burning in America's electric power-generating plants produces 80–85 percent of America's annual emissions of sulfur dioxide. The other 15–20 percent comes from oil refining, commercial ships, and smelting of sulfide ores. When the sulfur dioxide leaves the smokestacks hundreds of feet above the ground it interacts chemically with the moisture and oxygen in the air to produce sulfuric acid, a major component of "acid rain." The problem is worldwide. China, because of its massive use of coal, leads the world in sulfur emissions.[6]

How strong is the acid in acid rain? How does it compare, for example with Coca-Cola, lemon juice, or battery acid? The measuring stick used to determine the level of acidity is the pH scale (figure 7.1). The scale extends from 0 to 14. Values greater than 7 are basic; values less than 7 are acidic. The scale is logarithmic, meaning that each unit change is an increase or decrease of ten times. Hence, a value of 5 is ten times more acidic than a value of 6, pH 4 is 100 times more acidic that 6, and pH 3 is 1,000 times more acidic than 6. Although some acid rains and acid fogs (Los Angeles)

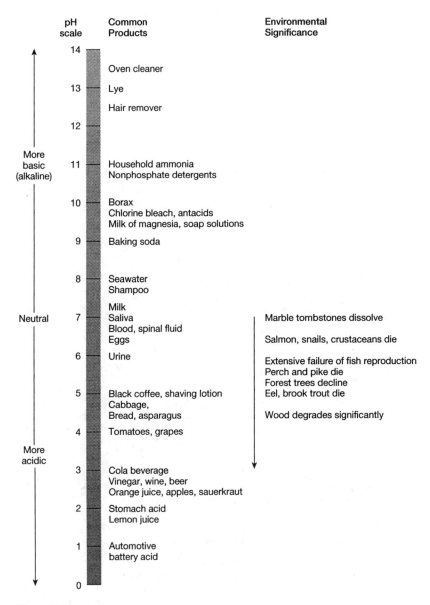

**Figure 7.1**
The pH scale, the pH of some common liquids and solids, and the environmental significance of acid pH values.

have been found to have the pH of lemon juice, average acid rain has a pH between 5 and 4, more acid than urine but less acid than tomatoes.

Unpolluted rainwater has a pH of 5.6, so it is acidic. When we speak of acid rain we mean rain more acidic (lower pH value) than 5.6, enhanced acidity rather than normal acidity (just as when we speak of the normal greenhouse effect versus the enhanced greenhouse effect). The acidity of rain over the Eastern United States, where most of our population and most of our coal-burning industries are located, is much lower than 5.6 (figure 7.2). Recall that each unit change is a change of ten times, so pH 4.6, typical of rain in the Eastern United States, is ten times more acidic than 5.6. Clearly, there is a problem with rainfall in the Eastern United States.

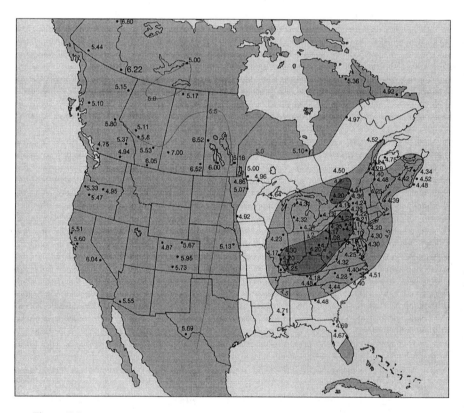

**Figure 7.2**
Contour map of pH values in the United States and Canada. Normal pH values dominate western areas, while a NE-SW-trending plume of more acid rain covers eastern United States and Canada (Environmental Protection Agency).

Well, so what difference does more acid rain make? There are several differences.

• Air with enhanced acid content kills fish in lakes. In a third of the lakes in New York's Adirondack Mountains the pH is commonly 4.3 and all the fish have died. Winds from the Upper Midwest blow toward the northeast, so that some of America's air pollution rains on eastern Canada. As a result, in Nova Scotia the pH has fallen so low in some rivers that salmon cannot live in them. Acid rain harms plant life and crops as well. It removes nutrients from soil, destroys leaves on trees, and mobilizes both nutrients and toxic metals in the soil, which end up in nearby lake waters.

• Acid precipitation in polluted cities causes statuary and building surfaces to deteriorate two to three times faster than in rural areas. Do you remember the supposed midnight ride of Paul Revere in 1775 to warn the American colonists that the British were coming? His tombstone doesn't. Acid precipitation has erased the inscription. The Parthenon, Taj Mahal, and Michelangelo's statues are dissolving under the onslaught of the acid pouring out of the skies.

• Half a century ago, if you stood on a hilltop on a clear day just about anywhere east of the Rocky Mountains, you could have seen things 70 miles away. Now, average visibility—even far from cities—is about 15 miles. There is a permanent haze. A 10-year study of visibility in 12 U.S. National Parks found continually decreasing visibility. According to the researchers the cause is mostly sulfur compounds. At times in the East "it is almost pure dilute sulfuric acid," according to one investigator. Between 1982 and 1992, summer sulfate hazes in the Great Smokey Mountains National Park, Tennessee, soared almost 40 percent. The most improved site was Chiracahua National Monument, Arizona. Sulfur levels dropped by one-third because of the closing of, or emission controls on, nearby copper smelters.

Fortunately, there has been progress in recent years in slowing the emission of sulfur from smokestacks, thanks to the Clean Air Act acid-rain amendments passed by Congress in 1990. Between 1981 and 2000 sulfur emissions dropped by 50 percent.[7] Industry was required to cut its sulfur emissions. In the first phase, which began in 1995, 445 power plants cut their sulfur emissions by 50 percent. In the second phase, which began in

2000, another 700 plants were required to start reducing their emissions. A 2003 study found that more than half of our artificially acidified lakes are bouncing back, their acidity reduced from previous lows,[8] but recovery is slow. The acidity took decades to develop and will take decades to reverse.

### Nitrogen Dioxide

Approximately 90 percent of nitrogen dioxide emissions (and those of other nitrogen compounds) are caused by human activities, about one-third from motor vehicles.[9] Thus, more than half the emissions of nitrogen dioxide are made at ground level, in contrast to sulfur emissions. Nitrogen emissions grew 4 percent between 1981 and 2000 and are one-third greater than sulfur emissions.[10] However, because nitrogen compounds are a major ingredient in the formation of smog (see below) near ground level, they are removed from the air very quickly near their source and are not a major cause of acid rain.

Until the mid-1990s, the catalytic converters in cars driven in the United States increased emissions of nitrogen oxides from the car. Since then, the catalyst in the converters has been changed and emissions of nitrogenous gases have been halted. The nitrogenous compounds are converted to nitrogen and water, Under ideal conditions, the converters can reduce nitrogen oxide emissions by 95 percent.[11] However, catalytic converters can only clean up vehicle exhausts if the engine is warmed up, and many trips taken by walk-avoiding Americans are too short or involve lots of stopping and starting, so the engine does not get hot enough for the catalytic converter to do its job.

### Ozone and Smog

As I will explain in chapter 8, ozone in the stratosphere (more than 10 miles above our heads) is essential for our well-being. But ozone near ground level is bad news. It is the chief component of urban smog and is most severe in areas with little wind, much sunlight, and many motor vehicles. Ground-level ozone is a secondary pollutant, because it does not come belching out of smokestacks or car exhausts. Most ozone forms when sunlight stimulates a chemical reaction between the nitrogen dioxide and hydrocarbon fumes coming out of car and truck exhausts. The main function of catalytic converters is to convert the fumes into carbon dioxide and water. In short, cars produce smog. In 2000 half of all Americans (more than 141 million) lived in communities that had severe smog pollution during the May-through-September period when higher temperatures and increased sunlight combine with stagnant air to produce ozone. The number of sufferers was up from 132 million in 1999. The air in nearly all the biggest cities, including New York, Los Angeles, Houston, Chicago, and Dallas, fell short of federal standards for ozone pollution.

Houston, because of its unusually high concentration of petrochemical plants, and the Los Angeles Basin, because of too many cars, have the nation's worst smog problems. With 10 million cars, Los Angeles has the highest ratio of cars to people in the world. Californians use 5 percent of the *world's* gasoline. Its 24,000 public school buses are among the most polluting in the nation. Examination of the lungs of young accident and homicide victims in Southern California found 75 percent had airspace inflammation and 27 percent had severe damage. That's one-quarter of *young* people. For these reasons, California is commonly at the forefront of the push for cleaner-burning gasoline-powered cars and hybrid cars.

Environmental efforts to reduce smog in the Los Angeles area have been paying off. In 1977 the number of stage 1 smog alerts, in which avoidance of vigorous outdoor exercise is recommended and those with health problems are advised to stay indoors, was 121. In 1996 there were only 7 such alerts. In 1999 the area had the best smog record ever recorded there. It was the first summer in 50 years without a single stage 1 warning. Sunburn advisories may soon become more common than smog alerts.

Smog is a particularly serious problem for asthma sufferers, who number 17 million in the United States, 6 percent of the population.[12] Every day 14 people die from asthma attacks. A study in France determined that asthma attacks increased by 30 percent on smoggy days. In Paris, 42 percent of visits to pediatricians are for respiratory problems. A study in the Netherlands found that people living near major roadways are twice as likely to die from heart and lung diseases as those who live farther away.[13] Smog in Europe kills more than twice as many people as car accidents.

Pregnant women exposed to the high concentrations of smog and carbon monoxide (see below) characteristic of large cities have triple the risk of having a child with heart malformations, the rate jumping from the normal two per thousand to six per thousand.[14]

## Carbon Monoxide

Carbon monoxide interferes with the ability of the blood to carry oxygen throughout your body. The hemoglobin in your blood is the oxygen transporter, but it likes carbon monoxide 200 times better than it likes oxygen. So it preferentially transports the monoxide to your organs instead of the oxygen your organs would rather have. Unfortunately, this gas is colorless and odorless, so it cannot be detected without special instruments. Symptoms of carbon monoxide poisoning include headache, nausea, vomiting, dizziness, coma, and even death. The amount of carbon monoxide in the air we breathe has risen over time in the Northern Hemisphere but the degree of increase is uncertain.

Guess where 60 percent of carbon monoxide pollution comes from? Why, motor vehicles of course. Between 1990 and 1999 carbon monoxide emissions decreased 2 percent and are at their lowest level since 1981.[15] Catalytic converters now present in most American cars remove the carbon monoxide from car exhausts (assuming the engine has warmed up), changing it into carbon dioxide, which won't kill you but instead contributes to global warming. Sometimes you win, sometimes you lose, and sometimes you just break even.

## Particulate Matter

It is much easier to remove the olive from a martini than it is to remove the vermouth. . . . We must rely on prevention rather than decontamination.
—Ivan L. Bennett, Jr.

There is a wide variety of microscopic particles floating around in the air we breathe. Some, like soil (dirt), have always been there and most of us are able to tolerate them if the amounts are not too great. But much microscopic airborne stuff today consists of human-made materials such as lead, rubber, soot from coal-burning power plants and motor-vehicle emissions, and poisonous chemical elements and compounds emitted from smokestacks. Some particulate materials, such as diesel soot, carry toxic compounds like benzene and dioxin that can increase cancer risk. The 15 percent of lung cancers not attributable to cigarette smoking are caused by particulate air pollution.[16] And it has been found that minuscule particles in the air pose a greater risk to the heart than to the lungs.[17] According to the EPA, particles in the air may be responsible for 60,000 deaths annually in the United States.

## Lead

The decline in atmospheric lead is perhaps the brightest spot on the air-pollution scene. Its dominant source was tetraethyl lead, formerly added routinely to gasoline to increase engine performance. Lead additions started in the mid-1920s and continued until 1986, when the sale of leaded gas was finally banned in the United States. Lead in the air has decreased 93 percent since 1981.[18] The blood of Americans now contains 50 percent less lead then it used to. Lake and reservoir waters have seen decreases of as much as 70 percent. Lead poisoning causes antisocial behavior and learning disabilities and may be one cause of criminal activity.

Because the major industrial countries have most of the cars and are phasing out leaded gasoline, the lead added to the world's gasoline dropped 75 percent between 1970 and 1993. But in Third World countries leaded gas still dominates, although there has been some progress. In 1997 the Chinese government banned the sale of leaded gasoline in Shanghai, one of the country's most polluted cities. Manila, capital of the Philippines, banned leaded gas in 2000. Many African countries have either banned leaded gas or are phasing it out. About half of the countries in Central and South America have introduced unleaded gasoline.

In China, two-thirds of the gasoline sold is leaded and studies from different parts of that country indicate lead poisoning of children living in industrial and heavy-traffic areas ranges between 65 and 99.5 percent, based on EPA standards. Perhaps 50 percent of children living outside such areas have lead poisoning. Of the 10 cities with the worst air pollution, 7 are in China. The microscopic lead particles emitted from motor vehicles can remain airborne for weeks          around the globe. Of course, the amount in the air           nce from the source increases and repeated rains wash          from the air.

## Rubber

Everyone who drives knows that tires wear and need to be replaced. Ever think about the microscopic pieces of rubber that are worn from the tire treads? Where do they go? Well, they don't disappear. They get into the air we breathe and adorn our lungs. Do your lungs need retreading? In addition to rubber particles, wear and tear of tires releases large amounts of chemicals called polycyclic aromatic hydrocarbons into the air and these are suspected of being carcinogenic.[19] About 15 percent of lung cancers are caused by factors other than cigarette smoking.

## The Particle-Size Controversy

What size of particle in the air is dangerous? Is it the larger ones or the smaller ones? Or perhaps it makes no difference. These questions are debated with great ferocity among industrial representatives, the federal government, and environmental groups. Until mid-1997 federal law mandated that the particles with diameters less than 10 micrometers (1/2,500 of an inch, one-seventh the diameter of a human hair) be below a certain threshold amount. Otherwise the air would be considered polluted with particulate matter and must be cleansed. The size of 10 micrometers was chosen by the EPA to limit regulation to those particles small enough to penetrate beyond the upper airways of your body's defenses (figure 7.3).

However, studies have indicated that the most damage to human lungs is done by particles less than 2.5 micrometers in diameter (1/10,000 of an inch, less than 1/40th the width of a human hair). These are the particles that penetrate deepest into the lungs and cause the most damage. They appear to be responsible for most of lung-cancer deaths among nonsmokers. So in 1997 the EPA changed the law to emphasize the amount of particles smaller than 2.5 micrometers in diameter. Industry objects to this change because most of these tiny particles result from the combustion of coal and oil (soot). The larger particles tend to be road dust and soil clay. The change from 10 to 2.5 micrometers cost industry a lot of money in new pollution controls.

Those opposed to the 2.5 standard also point out that the material smaller than 2.5 micrometers is so tiny and stays suspended in the air for so long before settling that much of it comes from sources outside any particular region. It is not within the control of those who would be penalized. For example, several states on the East Coast receive large amounts of this cryptodust from the Sahara Desert. Florida will be in noncompliance much

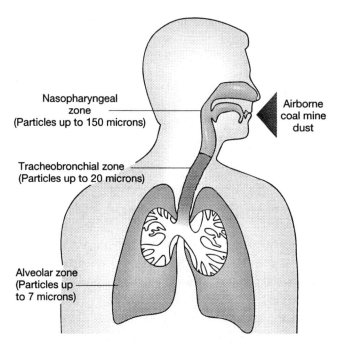

**Figure 7.3**
Schematic diagram of the upper respiratory system in humans showing the depth of penetration of different particle sizes of foreign objects that are inhaled. *Source: Earth and Mineral Resources*, vol. 59, no 7 (1990): 7.

of the time.[20] The Western United States receives cryptodust from the Gobi Desert in China. Critics also note that the size designation ignores the nature of the particle. The particles smaller than 2.5 micrometers in midsummer in the farm belt differ dramatically from those in the streets of a major city. One is mostly soil dirt, the other dangerous chemicals. The complexity of particle types is enormous but is not taken into account by considering size alone.[21] It is perhaps worth noting that 73 percent of the particles less than 2.5 micrometers in diameter have diameters less than 0.1 micrometers. These may be even more dangerous than their larger brothers. Should they be regulated? What price safety? The battle continues.

**Particle Pollution and Rainfall**
Recent research indicates that particles suspended in the air lower the amount of precipitation by preventing large droplets of water from forming.[22]

They have a dampening effect on rainfall, which can have effects that extend far beyond suppressing local precipitation. For example, tropical rain produces much of the energy needed for worldwide movement of air. Any change in rainfall in the tropics will affect global climate in uncertain ways. As environmentalists often note, everything is connected to everything else.

## What Price Clean Air?

When someone is chronically ill, the cost of pollution to him is almost infinite.
—Anonymous congressional staff member

Should the EPA disregard cost in establishing clean-air standards? A health-only approach is the historical norm in federal environmental policy. In recent years, however, talk of cost-benefit analysis has crept into most regulatory discussions. What is the average American willing to pay for what degree of health? Those who favor a cost-benefit approach say that a health-only standard is simply naive. If populations are going to be protected with a margin of safety, then *no* level of air pollution is acceptable. Some degree of balance is needed. A dollar's worth of benefit should be generated for a dollar's worth of expense. What are *you* willing to pay? Ten dollars a month for lowering the particle-size standard to 2.5 micrometers? How can you know how much health benefit the change from 10 to 2.5 micrometers will bring you? Tough questions.

## Planes and Pollution

As in a quiet backwater, pollution collects in the stratosphere and no rain washes it away.
—Louise Young, *Earth's Aura*

Commercial airlines are handling more passengers each year. Air traffic in the United States is expected to double between 1997 and 2017. Like cars, planes burn gasoline (jet fuel) and currently account for 10 percent of America's oil consumption. A jumbo jet burns 3,250 gallons *per hour* of flight; the now-retired Concorde, 6,500, spewing large amounts of pollutants into the atmosphere. And thousands of planes are flying every day of the year. Air transport is the fastest growing source of greenhouse-gas emissions, currently accounting for 3.5 percent of them. The newest and largest oceangoing ship, the Queen Mary II, burns 13,000 gallons of diesel fuel per hour. A car burns 2 gallons of refined petroleum per hour.

According to the Natural Resources Defense Council, a single 747 airliner landing and departing generates gaseous organic compounds equal to those of a car traveling 5,600 miles and nitrous oxide equivalent to a car driving 26,500 miles.[23] This suggests that people living downwind of large airports should suffer more health problems than other people. This expectation is borne out by a study of health records in Seattle. Those living downwind of the city's airport had significantly higher rates of respiratory diseases, pregnancy complications, infant mortality, and other health problems.

Nevertheless, airports are exempt from many air-pollution regulations.[24] In many cases they are among the top polluters in major metropolitan areas. The two major airports in New York City are among the ten largest sources of smog in the city. Los Angeles International Airport is second only to Chevron Corporation as a source of smog in the City of Angels. Chicago O'Hare International Airport is the fifth largest source of pollution in the Windy City area. The two airports in the Washington, D.C., area ranked between two garbage incinerators as the fourth and sixth largest sources of smog in the nation's capital. Clearly, airports need more pollution controls than they are now subjected to by the EPA.

In 1997 Zurich airport became the first in the world to charge stiffer landing fees for aircraft that emit more pollution. The cleanest aircraft pay 5 percent less than before, while the dirtiest pay 40 percent more. Is this a model the United States should adopt?

### Ships and Pollution

Man has lost the capacity to foresee and to forestall. He will end by destroying the earth.
—Albert Schweitzer

Commercial airlines are not the only polluting nonautomotive means of transport that is growing rapidly. Ocean-going ships are also of increasing concern.[25] During the past 15 years, as international trade has exploded and shipping capacity has grown by 50 percent, cargo ships have become one of the nation's leading sources of air pollution, threatening the health of millions of people living in port cities. These ships burn the dirtiest grades of fuel, literally the dregs of the oil barrel after refiners have removed cleaner fuels like gasoline and jet fuel. These low-grade hydrocarbons have the consistency of mud, with sulfur levels 3,000 times that of gasoline. They must be heated simply to allow them to move through pipes to enter the engine cylinders. A single cargo ship coming into New York harbor can release in an hour as much pollution as 350,000 2004-model cars.

Satellite photos show that trails of pollution thousands of miles long are causing semipermanent clouds above oceanic shipping routes. Scientists who study climate change are concerned about the effect on global warming as well as on atmospheric pollution. Foreign-flagged ships are responsible for almost 90 percent of the pollution in American ports. So far our government has refused to demand stricter regulations for these ships that dock at American ports.

### Progress In Cleaning the Air

Every American in every city in America will breathe clean air [by early in the next century].
—President George H. W. Bush, 1989

The marked decrease in air quality since the end of World War II provoked angry complaints from environmental groups in the 1960s, culminating in passage by Congress of the Clean Air Act in 1970 (amended in 1997 and 1990), despite vociferous opposition from industrial polluters. Within a few years air quality improved nationwide. Between 1970 and 2000 aggregate emissions of the six principal pollutants tracked nationally decreased 29 percent, despite a 45 percent increase in energy consumption.[26] Nevertheless, half of the people in the United States still breathe air that sometimes fails to meet the standards set by the Clean Air Act. Most of the problem is (ozone) smog generated by motor vehicles.

The greatest improvement in air quality has been in the amount of particulate lead. It has dropped like a, well, like a hunk of lead because of the phasing out of leaded gasoline. Sulfur dioxide levels have also shown marked improvement, although some of the decrease is illusory, resulting from an increased height of smokestacks in existing and new industrial facilities. Pollution measurements are made in cities, and the taller smokestacks simply transport their pollutant sulfur further away from the measuring stations. Clever, huh? The pollutant showing the least improvement is the nitrogen oxides that are an ingredient in ozone formation (smog), in large part because of the increase in motor-vehicle use.

## Indoor Air Pollution

Mother Nature says: "Clean up your room."
—Anonymous T-shirt slogan

Studies indicate that residents of highly developed countries such as the United States spend little time outdoors (table 7.1). This raises the question of air quality indoors, not only at home but also in the sealed office buildings where many American work. It is common for some air pollutants to be two to five times more concentrated inside homes than outdoors. Sources include air fresheners, hair sprays, and oven cleaners. At work the ventilation system in the building can be a source of both germs and noxious chemicals. For some people, the least healthy air they breathe all day is indoor air.

The pollutants that lurk indoors can come from a wide variety of sources, including cooking appliances, furnishings, household products,

**Table 7.1**
Average hours spent per day in various locations by adults in 44 U.S. cities

| Location | Employed men | Employed women | Homemakers |
|---|---|---|---|
| At home | 13.4 (55.8)* | 15.4 (64.2) | 20.5 (85.4) |
| At work | 6.7 (27.9) | 5.2 (21.7) | — (0) |
| In transit | 1.6 (6.7) | 1.3 (5.4) | 1.0 (4.2) |
| Outside | 0.7 (2.9) | 0.3 (1.3) | 0.4 (1.7) |
| Inside other structures | 1.6 (6.7) | 1.8 (7.5) | 2.1 (8.8) |

*J. M. Samet, M. S. Marbury, J. D. Spengler, *Health Effects and Sources of Indoor Air Pollution*, Part 1, p. 2. New York: American Lung Association, 1988.

and pets. Because modern, energy-efficient buildings tend to be tightly sealed, with very little fresh (?) air entering from outdoors, pollutants can reach high levels inside. Most of the time, however, air-pollutant levels indoors are low and have not been shown to pose a serious health threat for most Americans.

## Cigarette Smoke

The most common and serious indoor air hazard faced by Americans is one that is produced deliberately, secondhand cigarette smoke. People in a room with a smoker breathe in cigarette smoke that contains about 4,000 different chemicals, including more than 40 cancer-causing agents.[27] Among the chemicals in the smoke are arsenic, methanol (rocket fuel), and toluene (banned for use in nail polish), a bit of lead, formaldehyde (embalming fluid), and hydrogen cyanide (gas-chamber poison), naphthalene (used in moth-repellent balls), a smidgen of DDT (the classic pesticide), a bit of butane (the stuff we use to start fires quicker), cadmium (a heavy metal found in batteries, which is implicated in cancers that have struck down thousands of industrial workers exposed to it), and traces of the highly carcinogenic polonium-210 (a radioactive substance).

Continuous involuntary smoking increases your risk of lung cancer, high blood pressure, rheumatoid arthritis,[28] heart disease (by 20–70 percent),[29] respiratory infections, and other health maladies known to be associated with the cigarette smoker. The National Research Council and Surgeon General of the United States have both come out strongly against cigarette smoke, no matter how you encounter it. As stated by the World Health Organization in an unusual worldwide release, "Passive smoking does cause lung cancer. Don't let them [the cigarette companies] fool you."[30] As former Surgeon General Charles Everett Koop said, "The cigarette is the only product in the world that kills if you use it according to the manufacturer's instructions."

The risk of being affected by secondhand cigarette smoke is particularly high for children, house pets, spouses of smokers, pop musicians, and bartenders, people continually exposed to high levels of air pollution. The unborn children of pregnant women are affected by the chemicals in the mothers' blood, and children of smoking mothers have significantly lower IQ scores and are 27 percent more likely to have respiratory illnesses during their first six years.[31] Passive cigarette smoking damages your arteries about 40 percent as much as it does to the smokers themselves.[32] Women occasionally exposed to cigarette smoke are 58 percent more likely to develop coronary heart disease; those regularly exposed are 91 percent more likely.[33] Secondhand cigarette smoke is also implicated in cervical cancer[34] and hearing loss.[35] Anyone who has been to indoor music festivals or bars where smoking is permitted is well aware of the air pollution in such places. In a remarkable, precedent-setting decision, the U.S. Supreme Court ruled in February 2004 that an airline can be held liable for the death of a passenger from a severe asthma attack caused by exposure to secondhand smoke.[36]

On June 16, 2003, the World Health Assembly drafted a tobacco treaty that was signed that same day by 28 nations. The United States was not among them. The document calls on each country's politicians to pass laws protecting its citizens from exposure to tobacco smoke. The treaty requires each ratifying nation to ban all tobacco advertising, promotion, and sponsorship. I can only suggest that you hold your breath when in the presence of tobacco smoke, but don't try to hold it until the United States signs this treaty.

### Radon Gas

Radon is a colorless, tasteless, odorless, radioactive gas that is produced by the natural disintegration of uranium in rocks and soils. Although it originates outdoors, it is heavily diluted in the air and poses no threat. Radon gas is threatening only indoors, when it accumulates to high levels in poorly ventilated areas such as the basement of tightly sealed homes with cracked floors.

A few decades ago, one of those famous "faceless bureaucrats" in Washington established a "safe limit" for exposure to radioactivity from radon, based not on health considerations but on his estimate of what an average American could afford to pay for cleanup. His estimate of your financial resources for radon cleanup was $500–$2,500. Using this criterion, the EPA estimated that about 7 percent of American homes were at risk and that "radon is a national health problem." Incredible but true. Both the EPA and the Surgeon General recommend that all houses be tested for radon. They recommend taking remedial action if radon levels of more than 4 picocuries per liter of air are found. An average American home has 1.6 picocuries.[37] In other nations the average is much higher: Finland, 4.6; Sweden, 3.8; Norway, 2.2; France and Germany, 2.9; Denmark, 1.9.[38] Most of these nations have set higher levels because their uranium-bearing rocks are widespread and they feel that a level of only 4 picocuries is an unrealistic goal. Finland's recommended action level is 7.4.

Suppose the radon level in your basement exceeds the EPA's upper limit. Should you panic? It's a judgment call. Research indicates that people who have lived for 20 years in houses that have 27 picocuries of radon per cubic meter (35 cubic feet) face an additional 2–3 percent chance of contracting lung cancer.[39] But the average American moves 10–11 times during a lifetime. If only 7 percent of homes are in danger (1 home in 14), the chance of moving to another home that is over the radon limit is small. In summary, the radon threat is negligible for all but a very few people who may live their entire lives on rock or soil with an extremely high uranium content, and who have an unventilated basement in which they spend most of the day.

### Air Pollution in the Office

In July 1976, a faulty cooling tower allowed the pathogenic bacterium *Legionella* to be dispersed through the air-conditioning system in a Philadelphia hotel, causing 182 of the guests there to become ill with Legionnaires' disease. Before the cause of the sickness had been discovered, 29 people had died. This is perhaps the best-known example of how the indoor human environment, where we spend about three-quarters of the 168 hours of each week, can become deadly.

In the United States, up to 21 million employees are exposed to poor indoor air quality.[40] Several major office buildings have recently made headlines by being diagnosed as "sick." A new disease called "sick-building syndrome" (SBS) has arisen. The disease had its origin in 1973 when the energy crisis caused by the Arab oil embargo dictated a cut in air-handling costs. The standard for the minimum amount of outdoor air brought into buildings was reduced by 70 percent.[41]

The outdoor-air cutback was accompanied by a gradual rise in the use of photocopiers, laser printers, personal computers, and other equipment that may release chemical fumes. What's more, architectural designs changed, and sealed windows, wall-to-wall carpeting, and fiberglass or particle-board materials that may also contribute to the problem were increasingly used in buildings. The lower rate of air exchange (ventilation rate) combined with increased exposure to indoor pollutants probably explains the rise in SBS illnesses. Studies show that a doubling of ventilation rates does decrease SBS symptoms.

In addition to chemical irritants, microzoos can arise in building ventilation systems. Fungi, bacteria, viruses, algae, and other microbes lurk inside air ducts, grow around ceiling tiles, and thrive on almost any damp surface. Bacteria and fungi can produce airborne particles such as spores that leave employees with symptoms like coughing, headaches, and other allergic reactions. In exceptionally severe cases, biologically polluted air can lead to serious illness or even death, as in the case of Legionnaires' disease in 1976.

## Conclusion

Based on the lung diseases, heart problems, and cancers attributable to air pollution in the United States, the quality of the air Americans breathe is bad. According to the Harvard School of Public Health, air pollution causes 60,000 deaths each year.[42] But air quality has improved markedly (29 percent decrease in total pollutants) since passage of the Clean Air Act in 1970, over the prostrate bodies of the major polluters. The noxious materials in the air include sulfur dioxide, nitrogen dioxide, ozone, carbon monoxide, and soot. All of these materials arise from the combustion of coal and oil and cannot be eliminated without a major decrease in their use. However, improvements in catching the pollutants before they leave smokestacks and the exhaust pipes of motor vehicles have had a beneficial effect. An unchecked and rapidly growing source of air pollution is airplanes. They are currently exempt from most pollution rules, and this needs to be changed.

Indoor air is normally safe, although there have been some highly publicized exceptions. Few people are in serious danger from noxious chemicals or biological agents, either at home or where they work.

# 8

## Skin Cancer and the Ozone Hole

There was no moment when I yelled "Eureka!" I just came home one night and told my wife, "The work is going very well but it looks like the end of the world."
—Sherwood Rowland, on his discovery that CFCs damage the ozone layer

In chapter 7 we noted that ozone is the main component of the foul stuff known as smog, and that it forms at ground level by a chemical interaction between sunlight and the fumes coming out of car exhausts. However, not all ozone is bad. It depends on where it is in the atmosphere. High above our heads at an elevation of 10–22 miles there is a zone of air with a concentration of ozone six times that at sea level, and about ten times as much as the amount in the ground-level smog in a smoggy city such as Los Angeles or Houston. This zone is known as the ozone layer. Before 1985 no one but meteorologists was aware of the ozone layer. The average person was oblivious to its existence and, of course, they had every right to be. It was just another one of those obscure facts that might be useful on a TV game show but where else? It had no effect on daily life.

In recent years, however, this 12-mile thickness of air in the stratosphere has garnered perhaps millions of inches of newspaper space and untold hours of television time. And the word *ozone* is nearly always followed by the word *hole*. What is ozone, how can a gas have a hole in it, and why did it suddenly become so popular?

**What Is Ozone?**

Lucky is he who has been able to understand the causes of things.
—Virgil

Ozone is a molecule (a unit of matter) composed of three atoms of oxygen. This is in contrast to the oxygen we breathe and that is essential for our existence, which is a molecule composed of only two atoms of oxygen. The ozone is formed in the stratosphere by a reaction between oxygen and ultraviolet radiation from the sun. The ultraviolet radiation splits the two-atom oxygen molecule into separate atoms, which then recombine with two-atom oxygen molecules to form the three-atom ozone molecule.

However, ozone is an unstable molecule. Left to its own devices the molecule will decay back to an oxygen molecule and a single oxygen atom. This sequence of formation and decay maintains an equilibrium between the amounts of ozone and oxygen molecules. So long as solar radiation is constant, the proportions of ozone and oxygen in the ozone layer will be constant. The oxygen and ozone molecules are constantly passing the extra oxygen atom back and forth in a never-ending game of molecular table tennis. The equilibrium is such that the amount of ozone is small everywhere, but is enough to protect us from harmful radiation. Even in an undamaged ozone layer the amount of ozone is only about 10 parts per million (ppm). If all the ozone in the atmosphere were brought to sea level, the thickness of the ozone column would be only as thick as two pennies.

**Why Is Ozone Important?**

Damage already done to the ozone layer will be with us, our children and our grandchildren throughout the 21st century.
—Margaret Thatcher, former British Prime Minister

Ozone abundance may be small but it is mighty, and is essential for human existence. The ozone molecule absorbs ultraviolet (UV) radiation from the sun and thus forms a barrier that prevents it from reaching the earth below. About 6 percent of the sun's radiation reaches the earth as UV.[1] Ozone is a cosmic sunscreen with a very high sun protection factor (SPF). Ozone acts like the stuff you (hopefully) smear on yourself at the beach. UV radiation is very high-energy radiation capable of breaking chemical bonds between atoms in living organisms. It damages the DNA in the cells of your skin, an effect that is cumulative during your lifetime. Without the ozone layer, intense UV radiation would cause more skin cancer, damaged immune systems, eye cataracts, genetic mutations in our offspring, and poorer farm crops. Increased UV also disturbs the reproduction and growth of microscopic one-celled plants (phytoplankton) that float on the ocean surface. These plants are the foundation of the entire ocean food chain, so a large decline in their abundance would be disastrous for all animal life in the sea.

Crops and livestock are also victims of increased UV radiation. Research has shown that yields of various types of beans and peas decrease by 1 percent for each 1 percent increase in UV radiation. And cataracts and eye cancers are increasing in cattle. Larval stages of shrimp and fish are also harmed by UV radiation. If unchecked, the loss of our protective ozone shield has the potential to seriously affect the food supply. A healthy ozone layer is essential to healthy human life, or perhaps *any* life, on earth.

## What Have We Done to the Ozone Layer?

If you calculated the real cost of a can of aerosol hair spray [in terms of ozone-layer damage], it would be $12,000–13,000.
—Hazel Henderson

Humankind, in its never-ending refusal to leave well enough alone, has attacked the ozone layer. To be fair, the attack was not really deliberate but was a by-product of one of our creations. In 1928, industrial chemists developed compounds called chlorofluorocarbons (CFCs), very stable molecules with interesting physical and chemical properties. Within a short time, varieties of CFCs found many commercial uses. They could be used as refrigerants (Freon) in refrigerators and air conditioners, for cleaning electronic circuit boards, in the manufacture of foams used for insulation, in hair spray and insecticide cans as propellants, and in styrofoam drinking cups. CFCs were indeed valuable chemicals. In the United States, annual per capita use reached 2.4 pounds per person in 1985. The Freon in automobile air conditioners, found in 90 percent of all new cars sold in the United States, contributed 27 percent of all CFCs released into the environment.

Liquid CFCs evaporate easily and are released into the air. Normal vertical air circulation raises them into the atmosphere, and in about 5 years they reach the ozone layer, where the trouble begins. The CFCs are very stable molecules at ground level and even in the stratosphere, where they are bombarded by UV radiation and last for many decades. Some types of CFCs can last in the ozone layer for more than a hundred years.

Under bombardment by UV radiation from the sun, the CFCs decompose by releasing their chlorine atoms (the *chloro* in the name). The chlorine atoms attack and destroy ozone. Each chlorine atom can destroy 100,000 ozone molecules, so even a little chlorine goes a long way toward destruction of our protective ozone layer. There are now five times more chlorine atoms in the upper atmosphere than before CFCs were invented, not good news for life on earth.

The chlorine atoms in CFCs are responsible for about 90 percent of the ozone molecules that are destroyed in the ozone layer. The remaining 10 percent result from bromine atoms, an element similar to chlorine but not derived from CFCs. Anthropogenic bromine gets into the atmosphere from a bromine compound in agricultural pesticides. Bromine atoms are 100 times less common than chlorine but are 45 times more effective in destroying stratospheric ozone. Unlike CFCs, the bromine compound lasts less than 1 year in the atmosphere and will disappear shortly after we stop using it on crops.

Chlorine atoms are also released into the air in great volumes by volcanoes during volcanic eruptions. But the chlorine from this source is very soluble in rainwater, so it is washed from the air before it has a chance to rise 10 miles up into the ozone layer. It is unlike the chlorine in CFC molecules, which does not dissolve in water and therefore rises unhindered into the ozone layer.

## How Much Ozone Destruction Has There Been?

We have met the enemy, and it is us.
—Walt Kelly, *Pogo*

Measurement of global ozone began in 1931, and the data show that the amount of ozone has been in serious decline in recent years. The first recognition of loss was measured over Antarctica in 1977 and was repeated every year thereafter. Incredibly, the continuing decrease was dismissed as a product of instrument inaccuracy until 1987. During the 1987 measurements a record loss of 50 percent of the normal Antarctic ozone was found, accompanied by an exceptionally high concentration of a chlorine-containing molecule. This was the "smoking gun" that indicted CFCs as the cause of the decline. Record or near-record ozone depletion has been measured in the Antarctic every year thereafter. The depletion of ozone over the South Polar Region is what journalists have dubbed the "ozone hole." It is the geographic area where lots of ozone molecules have been destroyed.

Ozone concentration is measured in Dobson units, named after an early-twentieth-century meteorologist. Dobson units measure the volume of ozone in a vertical column of air at a particular location. In 1955 above Antarctica ozone measured 320 Dobson units; in 1975, 280; and by 1995 it had declined to 90, a drop of more than 70 percent from the protective 1955 level. At that rate of decline the ozone-rich layer above Antarctica would have disappeared by now. A "hole" is arbitrarily defined as an area where the volume of ozone is less than 220 Dobson units, a decrease of about one-third of the normal value, and the area included in the ozone hole since 1955 was growing. The thinned area is now much larger than Antarctica and extends over southern South America and New Zealand to latitude 40°S. As a result of the loss of some of their protective ozone shield, summertime levels of UV radiation in New Zealand have increased 12 percent between 1989 and 1999. In Punta Arenas, at the southern tip of Chile, latitude 53°, sunburn and skin cancer were virtually nonexistent a few decades ago. The expanding ozone hole changed all that. Since 1994, skin cancer has soared 66 percent. Since UV-related disorders take decades to surface, the true impact may not be known for many years.

Subsequently, thinning of the ozone layer was detected over the Arctic, where ozone loss averaged 25–30 percent between 1971 and 1997. Losses in some parts of the Arctic are now 60 percent. Ozone loss is most severe in the Polar Regions because the chemical reaction that destroys ozone is stimulated by very cold temperatures, temperatures below -110°F. Ozone loss is greater above the Antarctic than above the Arctic because temperatures high in the atmosphere are colder there, -144°F in the Antarctic compared to -108°F in the Arctic. Ozone loss in midlatitudes so far is only 6–7 percent.

What Can Be Done?

A conference is a gathering of important people who singly can do nothing but together can decide that nothing can be done.
—Fred Allen

When the scientific community pointed out the decline and its likely effects to political leaders, the nations of the world panicked. Even politicians can be mobilized when there is a supremely clear and present danger, and this was certainly it. World leaders seldom agree on anything, yet they saw the urgent need to halt the production of CFCs and other ozone-destroying chemicals and replace them with other (more expensive) chemicals (HFCs, which, unlike CFCs, have short residence times in the atmosphere). In 1987, but before the "smoking gun" that indicted CFCs was in hand, representatives of more than 30 industrial countries approved the Montreal Protocol to phase out CFCs. They operated under the "precautionary principle" of acting before you are sure. In 1990 they tightened it and in 1992 tightened it still more.

Most industrialized nations stopped producing CFCs in 1996, but Russia continued producing them until 2002. Russia claimed the delay was necessitated by a lack of money, and the richer nations refused to subsidize the cost. Developing nations have until 2010 to stop, and China and some other countries (plus Russia until recently) have used this advantage to increase production and to enhance international smuggling operations.[2] The thinning of the Antarctic ozone layer has at least slowed, if not stopped, but it will be at least 50 years before full recovery occurs.

There was also a push to ban methyl bromide, the bromine compound used in pesticides that is responsible for a small but significant part of ozone depletion. Developed countries are responsible for 80 percent of worldwide production (half of that by the United States),[3] and they have agreed to end production by 2005; developing countries have until 2015. The poorer countries have access to a global monetary fund to help them make the transition to using and producing alternatives to ozone-depleting chemicals. Almost $1.5 billion had been provided to the fund by the end of 1999. Ozone-destroying bromine compounds in the atmosphere peaked in 1998 and have since declined by about 5 percent.

The use of major ozone destroyers has now declined by 85 percent, and the abundance of ozone-eating chemicals in the stratosphere has peaked.

The maximum size of the ozone hole was reached in September 2000 at over 11 million square miles, more than three and a half times the size of the 48 conterminous United States. However, the ozone hole won't begin shrinking for at least another few years and won't recover fully until at least 2050. Between now and then there will be yearly ups and downs, the exact size of the hole varying with yearly vagaries of Antarctic atmospheric temperatures. But the trend will be a decrease in hole size.

## Montreal vs. Kyoto

Love of money is the root of all evil.
—1 Timothy 6:10

It is interesting to compare the difficulties in getting worldwide agreement on limiting emissions of greenhouse gases (chapter 6) with the speed at which agreement was reached to phase out ozone-depleting chemicals. It is apparent that the United States and a few other industrialized nations are not going to abide by the Kyoto treaty but have no serious problem with the Montreal CFC treaty. It is also noteworthy that the Kyoto treaty, even if implemented worldwide, would have only a negligible effect on slowing enhanced global warming, but the CFC treaty will essentially eliminate the ozone-destroying chemicals. What is the explanation for the difference in response between Kyoto and Montreal?

The answer, as usual, is money. A major reduction in emissions of greenhouse gases by the United States would require very large expenditures by industry to shift from coal-fired power plants to gas-fired ones. It would also seriously impact the automobile industry, which would have to either massively increase the fuel economy of its cars and SUVs, produce fewer cars, or rush into production hybrid vehicles before the general public is ready to accept them. In addition, the federal government would have to invest many billions of dollars to push forward the development of renewable sources of energy such as wind and solar power, probably over the prostrate bodies of the large and politically powerful fossil-fuel industry. In short, abiding by Kyoto would cause great disruptions in the way Americans are used to operating. And Kyoto's effect on enhanced greenhouse warming is very small.

The Montreal CFC treaty, in contrast, requires few changes in the way we do things. Only the producers of a single group of products, CFCs—mostly

large chemical companies who also produce the replacements for CFCs—
are affected. Banning CFCs to save the earth's plants and animals, including
us, brings excellent worldwide public relations at a very low cost.

## Ozone and Global Warming

The elements of the climatic system are tightly coupled and the system is nonlin-
ear (i.e., a small stimulus could cause a large response). Slight changes in one area
can set off feedback mechanisms that could cause an entirely unexpected result.
—Stephen H. Schneider

Many Americans are confused about the relationship between global
warming and the ozone hole. Are they related and, if so, how? It turns out
that global warming increases destruction of the ozone layer. It works like
this. The warming of the lower atmosphere because of an increasing concentration of the greenhouse gases (chapter 6) leads to cooling in the upper atmosphere, where the ozone layer is located.[4] Cold temperatures
increase the destruction of ozone molecules. Thus global warming increases ozone destruction. Before this interrelationship was recognized in
the late 1990s, it was estimated that regrowth of the ozone later would be
accomplished in about 50 years. Now it appears that it may take longer.

## Which Wavelengths are Harmful?

Radiation need not be feared, but it must be respected.
—Karl Morgan

Okay. So we know what the ozone layer is, why it is important, and that
we have caused it serious harm. Have climate scientists observed any effects so far on the biological world, particularly on humans? The answer
is a resounding yes. The most dramatic effect is a marked increase in skin
cancer. To understand why, we first need to say a few words about the
spectrum of ultraviolet light we receive from the sun (figure 8.1).

　　Radiation with wavelengths shorter than 0.01 micrometers is termed either gamma rays or X-rays. Radiation with wavelengths between 0.01 and
0.4 micrometers is called ultraviolet radiation and is further subdivided

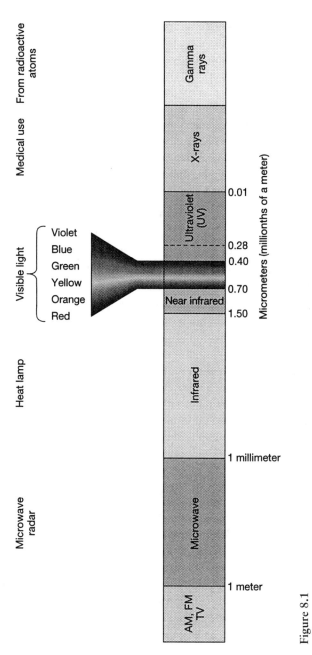

Figure 8.1
Electromagnetic spectrum of solar radiation. Most of the energy occurs in the wavelengths of visible light and the invisible near-infrared. Vision in animals and photosynthesis in plants are adapted to these wavelengths.

into UV-C, UV-B, and UV-A, with UV-C having the shortest wavelength and UV-A the longest. Like gamma rays and X-rays, ultraviolet rays are invisible to our eyes. Longer wavelengths, between 0.4 and 0.7 micrometers, are the ones our eyes are programmed to see as colors. Radiation with wavelengths longer than 0.7 micrometers is termed either infrared, microwave, or radio and TV waves, depending on wavelength. Our eyes cannot see these wavelengths.

The shorter the wavelength the more energy it has and the greater potential it has to damage living tissues. Wavelengths shorter than 0.28 micrometers (UV-C, X-rays, and gamma rays) are completely blocked by the oxygen atoms that make up 21 percent of our air. We know that wavelengths longer than 0.4 micrometers are not damaging to our eyes because we are bombarded continuously for 100 years or more without ill-effects. It is the wavelengths between 0.28 and 0.32 micrometers that may have the potential to harm us. These are the wavelengths that used to be stopped almost completely by the 320 Dobson units of ozone molecules in the undamaged ozone layer. However, as the ozone layer thins, more UV-B and UV-A pass through the atmosphere. The percent of UV-B radiation that passes through the atmosphere increases by about 1.6 percent for each ten-unit drop in Dobson units.[5] Not enough ozone molecules remain between 10 miles and 22 miles up to protect us effectively.

## Damage to People

Where there is no vision, the people perish.
—Proverbs 29:1

The human body is affected in several ways by increased UV radiation. Most noticeable are changes in the outer skin layers, an increase in skin cancer, and an increase in eye cataracts.

### Skin Cancer

The sun is the cause of at least 90 percent of all skin cancers.[6] We all know that the sun shines more intensely at low latitudes than at high latitudes. The tropics are hotter than the poles. Hence, we would expect, even with a healthy ozone layer, that skin cancer would be more common at low latitudes. And this is exactly what is found (figure 8.2). When the ozone layer

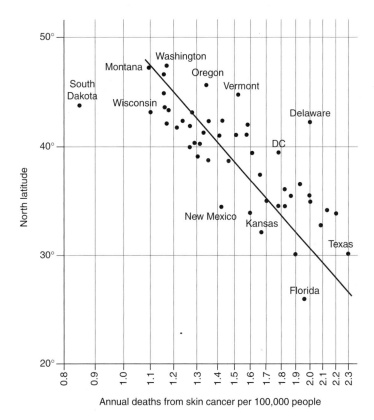

**Figure 8.2**
Annual deaths from skin cancer per 100,000 white males in the United States, compiled before thinning of the ozone layer. A latitudinal control is evident (Department of Health and Human Services).

was virtually undamaged in 1960, the occurrence of skin cancer among Caucasians increased by 10 percent for each 3° of latitude between 47° latitude and 30° latitude. The rate more than doubled if you moved from Montana to Texas.

About 1.8 million cases of skin cancer are diagnosed each year; 47,000 of them are fatal.[7] About one in seven Americans will contract this disease during their lifetime. A child born in 1999 has a one in five chance (20 percent!) of developing skin cancer later in life.[8] Almost all these cases will be due to overexposure to the sun's ultraviolet rays, most of which occurs before the age of 18. The main reason for the overexposure is the popular desire, particularly among young people, for a suntan, considered by our culture to improve physical appearance (the bronzed-god syndrome). The expression "healthy tan" is an oxymoron. A single blistering sunburn in a person 20–30 years old triples that person's chances of skin cancer. Young children are even more susceptible.

The form of skin cancer called melanoma spreads quickly and is the most lethal, killing 25 percent of those afflicted. If treated early the survival rate is more than 80 percent. In the United States the incidence of malignant melanoma is increasing at a rate of about 4 percent per year, a faster rate of growth than with the other, less dangerous, types of skin cancer (basal cell and squamous cell). It has more than tripled between 1980 and 2003, so that now about 5 percent of all skin cancers are melanomas.[9] The increase results in part from the continual thinning of the ozone layer and perhaps in part because of the increasing popularity of the seashore for recreation. There also is evidence that some people are genetically predisposed to get melanomas.

The chance of an American getting melanoma in 1980 was 1 in 250. Now it is 1 in 84.[10] In addition, the age at which melanoma appears is dropping. When first regularly reported, it affected persons aged 40 or older. By 1990 it was common in the 20–40 age group, which now accounts for 25 percent of cases. Melanoma is the most common cancer in women aged 25 to 29 and is second only to breast cancer in women 30–34. Nearly 60 percent of under–25 adults admit to "working on a tan."

### How Does It Happen?

The thinner the ozone layer, the more UV radiation you receive and the greater your likelihood of skin cancer. For each 1 percent loss of ozone molecules, skin cancer is expected to increase about 4 percent. If you lived near the South Pole, the heart of the ozone hole (where ozone concentrations have decreased 75 percent, from 350 Dobson units to 90), your skin-cancer risk would increase by 300 percent. The effect on the millions of resident penguins is unknown. Their numbers have already been seriously depleted by global warming and the resulting loss of pack ice, as well as by a decrease in their food supply—plankton and krill.

The National Weather Service publishes a *UV index* (table 8.1) that predicts the first two stages of skin damage from the ultraviolet rays. These

Table 8.1
Relationship between the UV index and minutes to burn

| Exposure category | Index value | Minutes to burn for most susceptible skin phototype | Minutes to burn for least susceptible skin phototype |
| --- | --- | --- | --- |
| Minimal | 0–2 | 30 | >120 |
| Low | 3 | 20 | 90 |
| | 4 | 15 | 75 |
| Moderate | 5 | 12 | 60 |
| | 6 | 10.5 | 50 |
| High | 7 | 8.5 | 40 |
| | 8 | 7.5 | 35 |
| | 9 | 7 | 33 |
| Very high | 10 | 6 | 30 |
| | 11 | 5.5 | 27 |
| | 12 | 5 | 25 |
| | 13 | <5 | 23 |
| | 14 | 4 | 21 |
| | 15 | <4 | 20 |

*Source:* National Weather Service.

are darkening of the skin as it tries to increase its ability to protect your internal organs by developing pigmentation that absorbs UV rays, and sunburn. The UV index covers 58 U.S. cities and is a forecast of the next day's ultraviolet intensity at the earth's surface when the sun is at its highest point.[11] It is valid for about a 30-mile radius from the city and, as with any forecast, local changes in cloud cover, haze, pollution, and other factors may alter the actual levels experienced. The index is designed so that the higher the rating, the greater the expected exposure to UV radiation and the faster a person is likely to sunburn. On an average summer day the UV index might be 6 or 7. At that index value, the time for even a less sensitive person to burn is generally less than an hour. For those with sensitive skin, skin burning starts in only 10 minutes.

Like other organisms habitually exposed to sunlight, humans have a layer of protective material to shield their internal organs. This protective material is called skin (figure 8.3). Human skin consists of three layers, called stratum corneum, epidermis, and dermis, from exterior to interior. The wavelengths that penetrate most deeply into your skin are located at

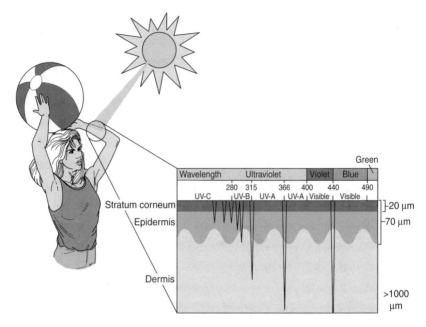

**Figure 8.3**
Cross-section of skin showing depth of penetration of various wavelengths. UV-B is considered more dangerous than UV-A because of its greater ability to break chemical bonds between atoms in your skin. *Source:* F. R. de Gruijl, "Impacts of a Projected Depletion of the Ozone Layer," *Consequences,* vol. 1, no. 2 (1995): p. 16.

0.440, 0.366, and 0.315 micrometers. Wavelengths in the visible range, such as the one at 0.44 micrometers, penetrate deeply into the dermis but pose a smaller health risk than those in the UV-A range at 0.366 and in the UV-B range at 0.32 micrometers. UV-B wavelengths are known to damage DNA, and there are several other penetrating UV-B wavelengths smaller than 0.32 that get through the stratum corneum and into the epidermis.

Probably the most noticeable impact of long-term overexposure to the sun is premature aging of the skin caused by UV-A radiation. Up to 90 percent of the skin deterioration we attribute to chronological aging is actually induced by ultraviolet radiation.[12] The UV rays damage the proteins that give the skin strength and resilience. The skin develops a leathery texture, wrinkles, and irregular pigmentation ("liver spots"). The higher the exposure to the sun, the younger the age at which these symptoms will develop. If you would like to know how young your skin would look had it not been repeatedly exposed to UV-A radiation, compare the skin on the outside of your arms with the side against your body.

Americans make about 25 million visits per year to tanning salons. Between 1 and 2 million Americans have a serious tanning habit, paying good money to cook in tanning salons as many as 100 times a year. These commercial tanning parlors are bad news. The object in these widely popular establishments is to produce a tanned skin in a short time. Hence, these parlors give off at least twice as much UV-A radiation as the sun does,[13] which does the tanning, with the result that their radiation is as bad as or worse than natural sunlight. Data indicate that artificial sunbathing increases the risk of eye damage and all types of skin cancer. Women who used tanning parlors once a month increase their risk of developing malignant melanoma by 55 percent.[14] An hour under the tanning light can be worse for your skin than an hour on the beach. "Beauty" can be damaging to your health and long-term survival. As the Skin Cancer Foundation puts it, "There is no such thing as a safe tan. A tan is the skin's response to an injury, and a sign of damage."

### Eye Damage

Your eyes lack a protective covering of skin but are less exposed to harmful radiation, shielded by eyebrows and eyelashes. Bushy and long are better than filmy and short. Women's fashions call for long and filmy, an inadvertent compromise with eye health. The depth of penetration of ultraviolet radiation into the eye varies with wavelength, with shorter wavelengths penetrating more deeply. Researchers estimate a 0.5 percent increase in cataracts for every 1 percent decrease in ozone concentration. Hence, the 6 percent reduction that has been measured in midlatitudes may result in a 3 percent increase early in the twenty-first century. Think of the penguins that have suffered a 75 percent reduction in ozone!

A cataract is a cloudiness that develops in the lens of the eye, the part of the eye that focuses an image on the retina at the back of the eye. The image becomes less clear as a result, requiring surgery if the cloudiness is severe. There are now 1 million cataract surgeries annually in the United States. If untreated, cataracts can cause partial or total blindness. When the lens in a camera is cloudy, your photos will be fuzzy. So it is also with the lens in your eye. Studies suggest that 20 percent of all cataracts are related to UV exposure.

In 1997 the U.S. Food and Drug Administration demanded that tanning salons instruct every customer to wear protective sunglasses. Unfortunately, closing the eyes, putting cotton wool over the eyelids, or even wearing sunglasses does not provide sufficient protection from the strength of the radiation used at tanning salons. The tanning rays used at the salons are 5 times more powerful than ordinary sunshine and hence do more damage to the eyes.[15]

## Sunscreens

In the eyes of Nature we are just another species in trouble.
—Lionel Tiger and Robin Fox, *The Imperial Animal*

Without a thinned ozone layer your chances of developing skin cancer are one in six. A decrease of 10 percent in thickness of the ozone layer could raise the risk at least 15 percent, but no one is certain. A significant increase in eye cataracts, which now afflict 10 percent of us, could also occur. How can we protect ourselves?

The most popular way is to apply sunscreen chemicals to your skin. Unfortunately, only one sunbather in three bothers with sunscreen lotion, which contains chemical compounds designed to absorb UV-B radiation. Some also absorb UV-A, which is desirable. Commercial sunscreen lotions are ranked by their sun protection factor (SPF), which applies to UV-B only.[16] SPF ratings indicate how much longer the sunscreen's use will allow you to be in the sun without getting sunburned. Say that skin begins to burn after 10 minutes of exposure. When it is protected by an SPF 15 sunscreen, a comparable burn will take 15 times as long. The higher the SPF the longer the protection lasts. Perspiration and water remove sunscreens, so they should be reapplied every 2 hours, even on cloudy days. Lotions with a sun protection factor of 15 block out 93 percent of the UV-B radiation; an SPF of 30 blocks out 97 percent. The American College of Dermatology recommends use of only lotions with an SPF of at least 15. In Ohio in 2003 the badly sunburned faces of three young siblings were enough to have the local sheriff charge the mother with the felony of "child endangerment" and throw her in jail. The felony charges were dropped a week later, but the message to the public was clear: neglecting to apply sunscreen to your kids is a form of child abuse.[17]

Hats with a brim at least 6 inches wide are also recommended, as are fabrics with a tight weave to keep out radiation. Sunglasses treated to absorb UV rays are also a good idea.

## Conclusion

Ozone is a gaseous molecule composed of three atoms of oxygen. It is concentrated in the "ozone layer" located from 10 to 20 miles above the ground and is our main shield against ultraviolet radiation from the sun. For the past 60–70 years, CFCs used in many common products have been drifting upward into the ozone layer and destroying ozone molecules, with the result that more UV-A and UV-B radiation has been reaching the earth's surface. Although ozone destruction is most severe in Polar Regions, it has now spread into temperate latitudes as well.

As a direct result of this increased ultraviolet radiation, skin-cancer rates have skyrocketed, particularly the deadly malignant melanomas. Authorities recommend that all people, and especially children, stay out of direct sunlight as much as possible and that exposed skin be protected with sunscreen lotions with an SPF of at least 15. Eye cataracts are also on the rise. Wide-brimmed hats and sunglasses that have been treated to protect against UV radiation should be worn in the summer.

Because the production of ozone-eating chemicals has, or is about to be, stopped, loss of stratospheric ozone has peaked. Recovery, however, will be slow but should be completed in about 50 years. The rapid response of the international community to the health threat caused by the thinning of the ozone layer is a fine example of what can be accomplished when cooperation prevails among nations.

# 9

## Nuclear-Waste Disposal: Not in My Backyard

Nuclear fission energy is safe only if a number of critical devices work as they should, if a number of people in key positions follow all their instructions, if there is no sabotage, no hijacking of the transport, if no nuclear fuel processing plant or repository anywhere in the world is situated in a region of riots or guerrilla activity, and no revolution or war—even a "conventional" one—takes place in these regions. No acts of God can be permitted.

—Hannes Alfven, Nobel Laureate, Physics

The central concern about nuclear power is the escape of radiation from the nuclear fuel. Cancer clusters have been found around nuclear plants worldwide. A U.S. government study found a high incidence of 22 different types of cancer at 14 different U.S. nuclear weapons facilities across the country.[1] There is evidence that working in a nuclear plant affects both the body cells and sperm of male employees. There are also data indicating that there is a far higher than normal occurrence of cancers in children living near nuclear power plants, and that the number of new cancers drops dramatically when the plants are shut down.[2] Added to this is the discovery of radioactive ants, roaches, rats, gnats, flies, worms, and pigeons near nuclear plants.[3] And there have been many thousands of minor accidents in reactors around the world,[4] 30,000 in the United States alone since the Three Mile Island accident 25 years ago,[5] as well as falsified safety reports at operating reactors. Compounding this has been the tendency of the federal government to either underestimate the dangers of radioactivity or to deliberately conceal relevant facts. And then there are major accidents such as at Three Mile Island in 1979 and Chernobyl in 1986. Is it any wonder that a large percentage of the American population fears nuclear power?

Fears have heightened since the 9/11 terrorist attack on the World Trade Center and the nation's capital. A scientific investigation completed in January 2003 reported that a successful terrorist attack on a spent-fuel storage pool at a large nuclear reactor could have consequences significantly worse than Chernobyl.[6] Because there is still no long-term storage site for nuclear fuel (see below), the risk will persist even if the reactors where it is now stored were shut down.

In this chapter we will deal with three topics. First, we will briefly discuss radioactivity and the principles of using uranium or plutonium to generate nuclear power. Next is the question of how to deal with obsolete and closed nuclear plants, our radioactive mausoleums. Finally, what do we do with the refuse the plants have already generated, mountains of radioactive waste, some of which will remain dangerous for millions of years?

## Radioactivity

Radiation need not be feared, but it must be respected.
—Karl Morgan

Uranium is the fuel used in nuclear power plants, and like all materials it is composed of atoms. An atom is the smallest unit of an element that still has the physical and chemical properties of that element. An atom of lead is simply the tiniest possible piece of lead. An atom is composed of one or more protons, neutrons, and electrons (figure 9.1). The protons and neutrons form the central core or nucleus of the atom. The electrons surround the nucleus and move around it in orbits, similar to the way planets move around the sun. Atoms are exceedingly small, an average diameter being about one hundred millionth of an inch. A piece of an element large enough to hold in your hand, such as a hunk of the element called lead, contains so many atoms that for practical purposes it can be considered infinite.

There are 90 different kinds of elements on earth and each is an atom that contains a different number of protons. The number of protons is the defining characteristic of an element; a different number of protons in an atomic nucleus means a different chemical element. The weight of an atom is the sum of its protons and neutrons, each counting one unit; the weight of an electron is negligible.

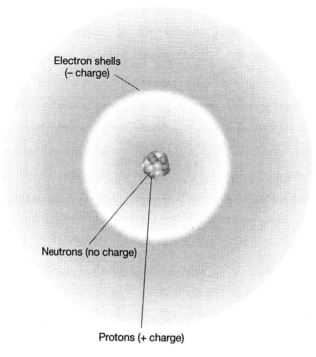

**Figure 9.1**
Idealized sketch of a lightweight atom showing arrangement of protons, neutrons, and electrons

In each of the 90 elements some of the atoms may be stable and others may be unstable. The unstable ones disintegrate by releasing particles and bursts of energy from their nucleus. These bursts are called radioactivity. The element called uranium is an example of an element that contains only unstable atoms. Each atom of uranium has 92 protons and, from the time Nature forms it, disintegrates in a series of steps, losing protons along the way, until it has only 82 protons and is an atom of lead. The complete transformation from uranium to lead in nature takes hundreds of million of years.

## Types of Radiation
Three types of radiation are given off during radioactive decay of an element. They are called alpha, beta, and gamma, the first three letters of the Greek alphabet. (Scientists like to do things with Greek and Latin.) An alpha particle is composed of two protons and two neutrons. It is very large, heavy, and

slow moving compared to beta and gamma emissions and hence has a very low penetrating power if it hits you. It can travel in air only 2 to 3 inches before being stopped by collisions with air molecules. It penetrates only two to three one-thousandths of an inch into your skin. Your body easily repairs the damage done to its outer covering by alpha-particle bombardment. If an element emitted only alpha particles you could safely carry around pieces of the radioactive element in your pocket. Uranium and plutonium are major radiators of alpha particles. However, they also emit beta and gamma radiation, so it is not advisable to play catch using a hunk of uranium.

All radioactive elements emit beta particles. A beta particle is an electron, a unit of negative electricity that is ejected from the nucleus of an atom as it disintegrates. They are about 100 times more penetrating than alphas, able to pass through one- to two-tenths of an inch of human skin. Although alpha and beta particles have a tough time getting very far into your skin, they have a much easier time with your internal organs. Internal tissues are not as dense as skin. If you inhale radioactive atoms you can suffer serious internal damage, perhaps cancer.

Gamma radiation is not particles but bursts of energy similar to visible light or X-rays. Because gamma rays have very short wavelengths, they have very high energies. Most radioactive atoms emit gamma rays. Their penetrating power is 10,000 times greater than that of alpha particles. Only thick shields of lead or concrete can stop gamma rays. Certainly your skin can't.

Summarizing, unstable radioactive elements such as uranium and plutonium are used to generate nuclear power. As they generate nuclear power their nuclei disintegrate, giving off three types of radiation, alpha, beta, and gamma. All three will damage your internal organs if they enter your body, but only gamma rays are powerful enough to pass through human skin. You must swallow or inhale radioactive substances to get alpha or beta rays into your body.

## How a Nuclear Power Plant Generates Power

Those who hope to keep our type of society intact and essentially unchanged while introducing so bizarre and dangerous a technology as atomic energy are flying in the face of ecological wisdom.

—Garrett Hardin, *Skeptic*

The fuel used in a nuclear reactor is a variety of uranium that has 92 protons and 143 neutrons in its nucleus (uranium-235). In the reactor the naturally very slow rate of disintegration of the uranium atoms is speeded up by bombarding them with neutrons. The resultant fast splitting (decay) of the atoms is called nuclear fission. There is so much energy released by nuclear fission that only 1 ounce of uranium produces the same amount of heat as burning 400 barrels of oil or 85 tons of high-grade coal. The heat boils water to generate steam, which turns a turbine to generate electricity.

In addition to releasing heat, the split nucleus also leaves atomic debris in the form of radioactive elements such as plutonium-239, strontium-90, radium-226, and radon-86. All of these are highly radioactive and dangerous.

## Lessons From Chernobyl

I'm not worried. Soviet radiation is the best in the world.
—Russian worker at the Chernobyl nuclear plant

The explosion of a nuclear reactor at 1:23 a.m. on April 26, 1986, at Chernobyl in *a remote area* in northern Ukraine, just south of the border with Belarus, was the worst accident in the history of nuclear power generation. The reactor experienced both a steam explosion and a partial meltdown of its nuclear core and released about 3 percent of its radioactive material.[7] The roof blew off the reactor building and enormous volumes of radioactive dust were hurled into the atmosphere to drift over the entire Northern Hemisphere. Worst hit were Ukraine, Belarus, and Scandinavia, although all of Europe was affected. In Germany, the teeth of children born after the Chernobyl explosion contain ten times the amount of radioactive strontium as the teeth of children born before the explosion.[8] In northeastern France, women's liver-cancer rates have gone up by 182 percent, lung cancer by 120 percent, and thyroid cancer by 283 percent in regions affected by Chernobyl fallout. Corresponding rates for men are 225 percent, 272 percent, and 86 percent. In France as a whole, thyroid cancers have doubled.[9]

The scattered radioactive debris was and still is a biological time bomb whose full effects will not be known for perhaps a hundred years or more. As of 2000, at least 9,000 people have died from the accident. In heavily contaminated areas there has been a 12 percent increase in birth defects. Inherited genetic damage has been found in children born in 1995 to parents who were exposed to the fallout. These birth defects include polydactyly (extra fingers and toes), and shortened limbs. The damage is to DNA in both sperm and eggs.[10] Such mutations become part of the genetic code and are passed down through the generations.

Ukraine and Belarus spend a substantial part of their national budgets dealing with health problems resulting from the Chernobyl fallout. There are 50 million people in Ukraine. As of 2000, 3.3 million of them (7 percent of the population!) have suffered illness as a result of the contamination, and the incidence of some types of cancer is 10 times the preexplosion average. According to the country's Deputy Health Minister in 2000, "The health of people affected by the Chernobyl accident is getting worse and worse every year." It is now 17 years since the disaster and the damage toll is still rising.

As of the end of 1998, 73,000 Ukrainians not yet killed by the disaster have been recognized by their government as being fully disabled by it, and another 323,000 adults and 1.1 million children are entitled to government aid for Chernobyl-related health problems. In 1996, 10 years after the disaster, Ukraine and Belarus still had 200 times more radiation in their affected areas than Hiroshima and Nagasaki had a decade after they were bombed to end World War II. The eventual death toll from the Chernobyl disaster will be in the millions over many decades.

The situation in now-bankrupt Belarus is equally horrifying. They received 70 percent of the radiation from the explosion in Ukraine, and the present state of the country is indicated by some population statistics.

Belarus is the same size as the United Kingdom but has only 10 million people.

Only 1 percent of the country is totally free of contamination.

One-quarter of all prime farmland is contaminated and permanently out of production.

Over 50,000 children have thyroid disorders. Nine hundred die each year from thyroid cancer.

As many as 800,000 children are at high risk of contracting cancer or leukemia.

Ninety percent of mothers nursing at the time of the explosion were giving breast milk that was radioactive.

Ninety-two percent of children suffer from something related to Chernobyl.

Twenty-five percent of the state budget is spent alleviating the after effects of Chernobyl.

Much worldwide publicity has attended the sharp increase in thyroid cancer among children in Ukraine and Belarus affected by the Chernobyl explosion. Noticeably increased rates of this disease occur more than 200 miles from the explosion; more than 50,000 children suffer from thyroid problems and about 3 percent of them have thyroid cancer,[11] a disease that occurs spontaneously in only one in a million children. The incidence is most severe among children who were under 4 years old at the time of the disaster; more than one-third of them are expected to develop thyroid cancer.[12] The average length of time for thyroid cancer to emerge is 17 years after exposure—that is, in 2003. The World Health Organization predicts at least another 50,000 cases of thyroid cancer in the coming years.[13] The explosion released large amounts of radioactive iodine-131, an element absorbed by the thyroid gland to manufacture thyroid hormone. Thyroid cancer is treatable and most of the children will survive, although their thyroid gland will need to be surgically removed and they will need thyroid-hormone replacements for the rest of their lives.

As a result of the Ukrainian experience with thyroid cancer, France—the country that depends most heavily on nuclear power—has handed out potassium iodide pills to 600,000 people living within 6 miles of a nuclear plant. The French government has about 2 million of the pills stockpiled in a warehouse just north of Paris. Austria and Switzerland also have iodine distribution programs. Swallowing these pills within a few hours of exposure to radioactive iodine can saturate the thyroid gland with non-radioactive iodine so it will not absorb the radioactive stuff. France's distribution program can only be viewed as political, since radioactive contamination spread more than a thousand miles from Chernobyl. Children living within 6 miles of a nuclear plant are hardly more at risk than those living 60 miles away. The World Health Organization has recommended that all schoolchildren in Europe have immediate access to potassium iodide pills in the event of a nuclear accident.

In 1998 the U.S. Nuclear Regulatory Commission said it would encourage states that house the nation's nuclear power plants to stockpile
potassium iodide pills. Late in 2001 the federal government bought 1.6
million of the pills and bought at least 6 million more in 2002. Following
the terrorist attack on the World Trade Center in 2001, California decided
to give potassium iodide pills to everyone living within 10 miles of a nuclear power plant. The nightmare continues.

The Minister of Agriculture in Ukraine, a grain-growing breadbasket
country similar to the American Midwest, said in 1996, "We know this
land should not be farmed, but if we don't . . . we will simply starve." It
will no doubt be many more decades before it will be safe to eat Ukrainian food. Grazing lands in faraway Scandinavia are still too radioactive to
raise cattle and reindeer for human consumption. In December 2002, 16
years after the disaster, an atomic food inspector in Moscow, about 420
miles north of Chernobyl, noted that "there are practically no cases of
radioactive watermelons this year." Now there's a minimum requirement
for you![14]

Organisms in the soil have also been affected by the radioactivity that
spewed from Chernobyl. Contaminated worms living within a few miles
of the accident are now having sex with each other instead of on their
own.[15] Two species have switched from asexual to sexual reproduction. It
is not known whether they will ever return to their former ways.

Those directly in the path of radiation from the Chernobyl explosion
were not the only ones affected by it. As of the end of 1998, 12 years after
the disaster, a total of 4,365 "liquidators," the name Ukraine gave to those
who took part in the Soviet cleanup effort, have died since 1986 of causes
"directly linked" to their work at Chernobyl.[16] In addition, endocrine disorders and stroke appear to be rising disproportionately among the
roughly 650,000 liquidators assigned to the initial "cleanup" efforts;
70,000 have been disabled by the radiation.[17] A recent study of liquidators' children conceived after the accident revealed they have seven times
as many mutations as did their older siblings, a sign that radiation had
damaged DNA in their parents' sperm and eggs. I would not want to be a
worker at this job site. The radiation levels inside the reactor building are
so high that plant employees joke that it's the biggest microwave in the
world.

The reactor's shattered fuel elements and other highly radioactive debris were entombed by the liquidators in a hastily constructed steel and concrete sarcophagus, which is already cracked and leaking radioactivity. In April 2003, Russia's atomic energy minister noted that the coffin "has a lot of holes" and that "we can see a situation where the roof could fall in, or other supports that hold up the roof could fall down. There is a strong chance it could happen."[18] Ukrainian officials agreed, noting that gaps in the shell totaled more than 10,500 square feet. They added: "The sarcophagus does not meet mechanical and structural safety requirements . . . thus there is a danger that parts of the shelter could indeed collapse. It is a realistic possibility."[19] Ukrainian and Western financial donors plan to build a new sarcophagus over the destroyed reactor. It should be completed in 2008. So far $11 billion have been spent on coping with the aftermath of the Chernobyl disaster.

The Russian-built reactors at Chernobyl are of a design not used in the United States, but 15 reactors of the same type are still in operation, 2 in Lithuania and 13 in Russia. Our reactors are considerably safer than those at Chernobyl. However, no one can guarantee that accidents caused by human error will not end in an American nuclear disaster. There have been a large number of serious nuclear calamities in recent years, the most newsworthy being the 1999 event at Tokiamura, Japan. Workers improperly handled uranium-235 and triggered a runaway chain reaction that burned uncontrolled for 20 hours. The International Atomic Energy Agency branded it the world's third worst nuclear accident behind the 1986 Chernobyl disaster and the 1979 event at Three Mile Island in Pennsylvania.

The Tokiamura, Japan, accident in 1999 unleashed radiation 20,000 times the normal level and injured at least 49 people, some critically. Local people were exposed to radiation levels estimated to be 100 times the annual safe limit. The effects of the radioactive rain that fell on the surrounding area have yet to be determined. Investigators have also found higher than expected uranium concentrations in the site environment, possibly indicating prior, unreported accidents. The Sumitomo Metal Mining Company was found guilty in 2003 of causing the Tokiamura accident and was fined $8,300. It voluntarily closed its nuclear-fuel-processing business. Six employees found guilty of causing the accident were given suspended prison sentences.

Despite the Tokiamura incident, the Japanese government seems inconsistent in its concern about the safety of its reactors. On the one hand, Tokyo Electric Power (Tepko), the world's biggest private electricity supplier, has closed 17 nuclear power plants because of safety concerns.[20] On the other hand, in a nuclear reactor at Shika on the northern peninsula of Ishikawa, 249 cracks have been found in a reactor shroud containing the main coolant and a small amount of water has leaked out. The Japanese nuclear regulator says no repairs are needed.[21] Perhaps more frightening was the admission by two power companies that they had neglected to report signs of reactor cracks first noticed in September 1998. And in August 2002 General Electric International, which built and maintains many of Tepko's plants, admitted that it had falsified safety records at 37 locations.[22] What do you believe should be the punishment for such dangerous and unconscionable behavior?

Cracks have also been found in the nozzles on the lids of 14 reactors in the United States. The lids are being replaced at a cost of $25 million each.[23] In April 2003, leaks were found at the bottom of a Texas reactor's pressure vessel, a site much harder to repair than leaks in the lid. It's getting harder to sleep at night.

Eastern Europe and the countries of the former Soviet Union contain 57 operating nuclear reactors that are not safe enough to be licensed in the United States. The countries housing these unsafe reactors are aware of their unsafe nature but do not have the money to do anything about it. The wealthier nations must do most of the paying to improve these reactors because, as Chernobyl demonstrated, *a nuclear accident anywhere is a nuclear accident everywhere.*

### Old and Obsolete Nuclear Power Plants

[The retired nuclear plants and wastes remain] a brontosaurus that has had its spinal cord cut but is so big and has so many ganglia near the tail that it can keep thrashing around for years not knowing it's dead.
—Amory Lovins, environmentalist

Originally, nuclear power reactors were given 30-year licenses by the Atomic Act of 1954, but the Nuclear Regulatory Commission in 1982 extended the operating license for commercial nuclear power plants to 40 years, beginning on the date of issuance. The location and year of expiration of the nation's 103 units is shown in figures 9.2a and 9.2b. The beginning date of 2002 reflects the reactors that came on line in 1962. Figure 9.2b does not take into account the possibility of license extensions. As of July 2003, 16 commercial nuclear plants had been granted 20-year extensions, and many others have signaled that they will apply soon. In 2001 President Bush, concerned about an impending energy shortage in the United States, indicated that he plans to approve license extensions to as many nuclear plants as possible, consistent with safety considerations.

When nuclear power plants were first built many decades ago no one foresaw them running for more than 40 years. The effects of intense radioactive bombardment, especially on metals, was seen as limiting the operating life of nuclear plants. And then there is the standard deterioration that occurs when any machine gets old. Robert Alvarez, senior policy advisor at the Department of Energy from 1993 to 1999, points out that nuclear reactors "are just like old machines, but they are ultra-hazardous." By pushing their operating span to 60 years, he says, "disaster is being invited."

The useful life of a nuclear reactor is uncertain. Nearly 100 reactors around the world have been closed down for various reasons since the era of nuclear power began about 40 years ago. Another 250 are scheduled for retirement within the next 10 years. Sixty-four of these reactors were more than 20 years old in 2000. How do you dismantle an old and obsolete nuclear reactor? The guts of the reactor are highly radioactive and will remain so for hundreds of thousands of years, longer than any civilization has lasted. The metal parts and the concrete used in construction have become brittle from long exposure to radioactivity. We cannot let the buildings crumble on their own, nor can we attack them with a wrecking ball or explosives as we would an old office building.

No retired reactor has yet been dismantled and no one knows whether such a thing is even technically feasible, given the level of radioactivity that must be dealt with. The incredibly radioactive core of each reactor must be emptied of spent fuel, and all pipes must be drained. All radioactive material, both solid and liquid, must be sent to a permanent waste-storage site. Then the building that houses the reactor must be disassembled and

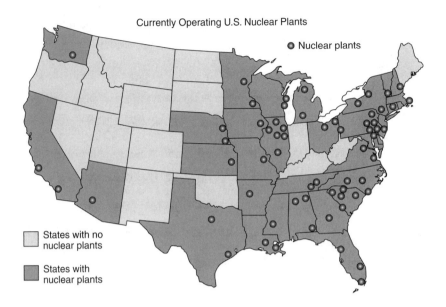

**Figure 9.2a**
Location of nuclear power reactors in the United States. Many sites have more than one reactor. There are no nuclear plants in Alaska or Hawaii (Nuclear Regulatory Commission).

hauled away to the same site. The companies that undertake this task will need to have mountains of insurance coverage against radioactive contamination for its employees. Possibly the federal government will have to be the insurer. I would not want to be one of the workers on this job.

Cost estimates for dismantling are uncertain but are at least $500 million for each of our 103 operating reactors, a total of $50 billion. The cost of cleaning up the nuclear-weapons complexes run by the military is estimated to be perhaps ten times this amount. No politician has yet wanted to tackle this problem. Meanwhile, the number of retired reactors continues to increase and their containment buildings are getting older and less structurally sound every year. No one has yet found a solution and the clock is ticking.

Aging commercial nuclear reactors are not the only problem America faces. The U.S. government's nuclear-weapons program has an unparalleled array of festering technical and environmental problems. With no easy way to solve them, the government in 1999 hired a company owned

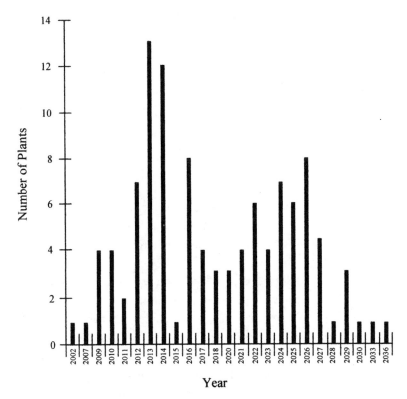

**Figure 9.2b**
Number of U.S. nuclear power plant licenses expiring between 2002 and 2036 (Nuclear Regulatory Commission).

by the British government to deal with it. How they propose do this remains to be seen, and debated, and debated, and no doubt debated some more. In 2000, estimates of the cost of cleaning up the contamination from nuclear-weapons production are 44 percent higher than estimates made only 2 years earlier, a colossal $200 billion. That's more than $700 to be paid by each man, woman, and child in the United States. Large families are getting more expensive every day.

## Storage of Obsolete Radioactive Materials

Hell no, we won't glow.
—Anonymous antinuclear slogan

The operation of America's nuclear power plants over the past 40 years has left us with another monstrous problem, and this one is still growing. What can we do with the mountains of nuclear-waste products that America's nuclear research, weapons program, and power plants have generated? Some of it will remain dangerous and potentially lethal for thousands to millions of years—that is, forever as far as the human species is concerned. More than half of all Americans live within 75 miles of an above-ground waste storage site.[24]

**Low-Level Waste**
Radioactive waste is classified for convenience as either low-level or high-level. Ninety-nine percent of all radioactive waste is low-level, meaning it will remain significantly radioactive and dangerous for *only* 300–500 years. In these wastes the maximum level of radioactivity is up to 1,000 times the amount considered acceptable in the environment. About 63 percent of this waste (but 94 percent of the radioactivity) originates in nuclear power plants. The rest comes from hospitals (cancer treatment), university and industrial laboratories (research), manufacturing (measurement), and military facilities (nuclear-weapons manufacturing and research). The range of materials in low-level waste is wide, and includes fabric, metal, plastic, glass, paper, wood, and animal remains.

Low-level wastes are usually sealed in metal drums, commonly after being burned in special incinerators to reduce their volume, and either stored aboveground in vast "holding pens," or buried in shallow trenches beneath about 3 feet of soil. In principle, the storage site should be carefully chosen with respect to the location and geology to ensure that there is no contamination of plants, soils, and groundwater. In practice, the metal drums corrode over time, and stress is placed on them by the continuous radiation from the material in the drums. Many drums at some sites have leaked radioactive liquids into the soil and groundwater.

Clearly, a safer method for storage of these drums is needed. The safest method is to store them in excavations tens of feet deep but far above the groundwater table in arid or semiarid areas of the Western United States. This method would keep water from the drums, and the waste could be easily and continuously monitored for a few centuries until it was no longer hazardous and could be landfilled.

Under congressional mandate, regional compacts that include several adjacent states have been established to store low-level waste on a regional rather than a local basis. There will be one storage site for each compact except for the Rocky Mountain compact, which will use the site of the Northwest compact at Hanford, Washington. Ten such compacts have been established, but only two compacts have settled on a storage site. However, lawsuits have been filed by environmental groups against some of the host states, and some compact states have recently withdrawn from their compact because of disagreements with other states in the compact. The future of the compacts is uncertain. Most low-level waste is still stored at the places where it was produced.

Another disposal method for low-level waste has been to cast it into concrete, encased in steel drums, and dump it into the deep ocean. Tens of thousands of tons of such wastes have been dumped at an internationally agreed site in the Atlantic Ocean 1,300 miles southwest of England. The concrete ensures that the waste will reach the floor of the ocean 15,000 feet down intact, where it is supposed to remain safe and undisturbed for at least hundreds of years. The concept is that any leakage over time would be diluted by the vast mass of ocean water and dispersed harmlessly into the surroundings. The problem with this approach is that as the drums begin to leak, a process that may already have begun, the fish that swim nearby become irradiated. When the smaller fish are eaten by larger fish the effects of the irradiation are spread in an ever-widening circle whose end is unclear. But it cannot be good. The best we can hope for is an insignificant effect. Our experiments with nuclear-waste disposal continue.

## The Waste Isolation Pilot Plant

Early on the morning of March 26, 1999, 45 years after the site search began, 25 years after initial studies started at the site, and ten years after its planned opening date, the first truckload of low-level waste arrived at the Waste Isolation Pilot Plant (WIPP) 25 miles east of Carlsbad, New Mexico. WIPP is the storage site for the federal government's low-level radioactive waste from 23 military sites across the nation and is located 2,150 feet underground within a huge mass of rock salt, the same stuff as the salt on your dinner table. Waste will be stored in a series of rooms about 325 feet long and the shafts will then be sealed with 13 different layers of concrete and earthen fills. In response to excavations, salt slowly creeps, much like glacial ice, and over 75 to 200 years will completely seal the mine opening. It is estimated that the radioactive waste will remain secure for more than 10,000 years. The repository has cost $2 billion for feasibility studies, construction, and defense against court challenges, which are not over yet.

The establishment of WIPP is unrelated to the site-selection process for commercial waste that has resulted in the nuclear-waste compacts described earlier. In what must rank as a masterpiece of understatement, Energy Secretary Bill Richardson said, "We won the Cold War by building nuclear weapons, but we have not cleaned up the legacy of the Cold War and its waste."

Over the next 35 years, barring successful court challenges, almost 40,000 truckloads of waste will be taken to the plant. The underground maze of 55 storage chambers at the site has a capacity of 6 million cubic feet of waste, only about half of the military's projected needs. By 2010, storage at WIPP will reduce the number of people living within 50 miles of an existing military nuclear-waste storage site from the present 61 million to only 4 million.

### High-Level Waste

High-level radioactive waste is a by-product of nuclear-weapons production and commercial reactors. It consists of spent fuel rods and liquid materials involved in reprocessing spent fuel to recover highly radioactive plutonium, and was produced at more than 100 former U.S. nuclear-weapons sites. High-level waste will be dangerously radioactive forever— that is, for at least tens of thousands of years. One estimate of their toxicity suggests that less than 1 gallon of the waste would be enough to bring every person in the world to the danger level for radiation exposure if it were evenly distributed. There are now about 100 million gallons in storage at 158 sites in 40 states (figure 9.3). Sweden, Finland, Japan, and Switzerland are the only countries that have succeeded in having repository sites for high-level waste accepted by their citizens. The United States is in the final stages of establishing a site.

In 1982 Congress required the federal government to find a suitable site for storing the nation's ever-growing accumulation of high-level nuclear

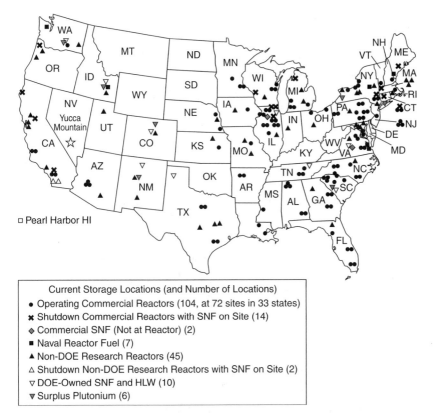

Current Storage Locations (and Number of Locations)
- Operating Commercial Reactors (104, at 72 sites in 33 states)
- ✖ Shutdown Commercial Reactors with SNF on Site (14)
- ◆ Commercial SNF (Not at Reactor) (2)
- ■ Naval Reactor Fuel (7)
- ▲ Non-DOE Research Reactors (45)
- △ Shutdown Non-DOE Research Reactors with SNF on Site (2)
- ▽ DOE-Owned SNF and HLW (10)
- ▼ Surplus Plutonium (6)

**Figure 9.3**
Approximate locations of surface storage sites for spent nuclear fuel (SNF) from commercial reactors and for other high-level waste (HLW) and radioactive materials (Department of Energy).

waste. In 1987 the Department of Energy settled on Yucca Mountain, 87 miles northwest of Las Vegas, Nevada. Feasibility studies have cost $4 billion.[25] As of 2001, the facility itself is projected to cost $57.5 billion, up 26 percent from the estimate made by the Department of Energy only 3 years earlier.[26] Movement of high-level waste from the present 131 storage sites in 39 states to Yucca Mountain was set to begin in 1985. But because of lawsuits by the State of Nevada and other groups opposed to the site, the opening was pushed back to 1989, then 1998, then 2003, and as of this writing the date is 2010. Few would bet on 2010 being the final date. Current (2002) Nevada governor Kenny Guinn said of the federal government's decision to use Yucca Mountain, "This decision stinks, the whole process stinks and we'll see him in court." With many lawsuits pending, Nevada state official Robert Loux said, "I think we can keep them out of the site for decades."

In the meantime, high-level waste continues to accumulate at the nation's reactors. Companies claim that their storage areas are full or nearly full and demand immediate action from the federal government. The current terrorist threat has increased the urgency of their demands. To transport the high-level waste to Yucca Mountain, more than 150,000 shipments of vitrified waste through 45 states would be made over 30 years, 3,000–4,000 shipments per year.[27] That's about 10 shipments per day, 365 days a year, for 30 years. Transportation would be by road, rail, and canal, and would pass through or close to many highly populated regions. This effort has been dubbed "Mobile Chernobyl" by the media and Yucca Mountain's opponents, a catchy but inaccurate phrase that sends shudders down a sane person's spine. Since 9/11, Americans have become aware that trucks and trains carrying nuclear waste are tempting terrorist targets. Of course, the transportation problem will be similar for any storage site. It brings to mind the well-known expression about being caught between a rock and a hard place.

The nuclear industry projects it will produce more than the Yucca Mountain capacity of 77,000 tons before 2020. This is an estimate for the waste produced by industry and does not include storage space for 2,500 tons of waste from military reactors and untold tons of glassified radioactive liquid wastes stored at former bomb plants. There has been no search for a second site, despite the fact that the first site is already oversubscribed. The federal government is still fighting with opponents of the first site. At present, high-level waste is stored "temporarily" at 130 nuclear power plants and military installations.[28] Eventually it will have to be moved *somewhere.*

The objections to storing high-level radioactive waste at Yucca Mountain are both political (NIMBY, Not In My Backyard) and scientific. The political objections stem from the fact that Yucca Mountain was not selected on technical grounds in the first place. Rather it was unfortunate enough to lose a game of musical chairs in the mid-1980s, when all of the other site options managed to find grounds (some technical, but usually political) to have themselves ruled out of the running. In 2004, however, Las Vegas, the nearest large city to Yucca Mountain, is one of the fastest-growing cities in America. Its powerful gambling industry will fight tooth and nail to oppose a project that it thinks has the potential to slow the hitherto insatiable flow of tourists to the gambling palaces.

The scientific objections stem from uncertainties about the safety of the facility from a geological point of view. Some scientists believe that serious earthquakes or volcanic eruptions may occur near Yucca Mountain within the next 10,000 years (figure 9.4). A magnitude 5.2 quake struck near Yucca Mountain in 1992. Another concern is that water can leak into the storage rooms 1,000 feet below the surface and cause the radioactive canisters to corrode. The Department of Energy's design calls for dense packing of canisters, which will result in radioactively generated temperatures of about 400°F, vaporizing any water in the rock envelope around the canisters and making it more mobile.

Related to the question of water leakage is the uncertainty caused by possible climate changes, irrespective of possible human-induced global warming. Climates can change very rapidly for reasons that are not now understood. Fifteen thousand years ago much of the Western United States was rainy, had high water tables, and contained many lakes (figure 9.5). The largest was 1,000 feet deep and was as large as Lake Michigan. Much of the Sahara Desert was fertile just 5,000 years ago. The prediction of future climate regimes is inherently risky.

A final decision on Yucca Mountain was made by the federal government in 2002. The Department of Energy made a final recommendation, which was then approved by both Congress and the president. Lawsuits to overturn this decision have already begun.

### Hanford—The Most Polluted Place on Earth

It is obvious that many projects justified by cost-benefit analysis do result in the predictable loss of life.
—Herman E. Daly, *Steady-State Economics*

You may now be thinking, as I did, about the most radioactively polluted place on earth. Where is it and how dangerous is it? The answer to this question is the Hanford site, 586 square miles in size, near Richland, Washington, where two-thirds of America's high-level nuclear waste is

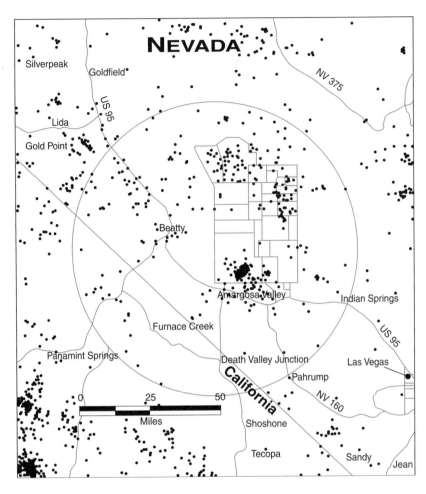

**Figure 9.4**
Location of earthquakes of magnitude 2.5 and greater from 1976 to 1996 in the
vicinity of Yucca Mountain. The cluster of earthquakes just to the southeast of
Yucca Mountain represents the main quake and 2,000 aftershocks from a 1992
quake (*Council of the National Seismic System Composite Catalog*, 1976 to
present, Southern Great Basin Seismic Network).

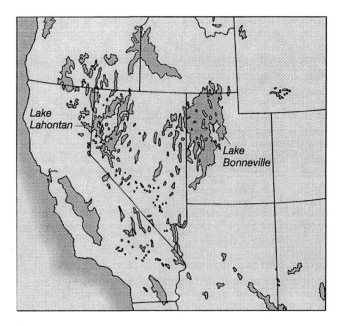

**Figure 9.5**
Location of large lakes in Western United States 15,000 years ago. The Great Salt Lake in Utah is the only large remnant still present.

stored. Hanford was established in 1943 to create plutonium for nuclear warheads. Plutonium does not occur naturally but is produced in nuclear reactors as a by-product of the fission of uranium. The reactors operated for 46 years, from 1943 to 1989, and produced 72 tons of plutonium. During its defense-production years the Hanford site was under strict military security and never subject to outside oversight. Former workers at Hanford (and other nuclear installations) are now known to have a significantly greater chance of getting cancer of the central nervous system, female reproductive system, and diseases of the thyroid gland than other Americans.[29] Sounds like a micro-Chernobyl without an explosion.

As a result of improper disposal methods, such as dumping 440 billion gallons of radioactive and other hazardous liquids directly onto the ground, Hanford's 586 square miles is still considered "the dirtiest place on earth." Four hundred and forty billion gallons is enough liquid to fill a lake the size of Manhattan Island to a depth of 80 feet. Anyone care for a swim? The volume of contaminated soil is about 1 cubic mile. In 2000 the Energy Department determined that the amount of plutonium and other human-made radioactive elements released into soil during nuclear-weapons manufacture was 10 times greater than it had estimated in 1987.

Hanford contains over 1,900 contaminated liquid- and solid-waste sites; more than 500 of these are located within 2,500 feet of the Columbia River, the Northwest's major waterway. There are 53 million gallons of highly radioactive waste stored in 177 single-shell underground tanks, 67 of which leak. So far, 1 million gallons of leaked waste have reached groundwater and have contaminated 270 billion gallons of groundwater over at least 100 square miles.[30] As the groundwater moves, so does the radioactivity, which has now reached the Columbia River, 10 miles from Hanford. Is it safe to eat the few remaining salmon that swim in the river?

In 1997 there were two chemical explosions at Hanford, which released some airborne plutonium. The site also stores 25 tons of solid plutonium that must be kept under constant armed guard, because only a tiny amount of plutonium is enough to manufacture a nuclear bomb. There are lots of terrorists and many rogue nations these days that would like only a few pounds from this 25-ton stockpile.

The DOE has described Hanford as "the single largest environmental and health risk in the Nation." Cleanup by 11,000 workers, some of them already getting sick from on-the-job radiation, is costing $2 billion a year, and the DOE estimates that repairing the environmental damage at Hanford could cost a trillion dollars ($1,000,000,000,000) and take the rest of the twenty-first century. That's an amount of money that would make even Bill Gates shudder. Engineers who have studied the site describe it as "the largest civil works project in history, it has no end." Even at the conclusion of the cleanup, Hanford will not be restored to a pristine, green field. It is hopelessly contaminated and will be fenced off to protect the public. It is essentially uncleanable and will remain hazardous forever. A report from the nonprofit Resources for the Future says that such permanent hazards will remain at two-thirds of the contaminated sites from the nuclear-weapons complex no matter how much is spent to remediate them.

## Conclusion

The fuel used for nuclear power generation is uranium. In a nuclear reactor the uranium is bombarded by neutrons, which causes the uranium atoms to split and release large amounts of energy that is used to generate electricity. Nuclear power plants are inherently very dangerous, partly because of the high levels of radioactivity they produce but also because of impossibility of eliminating human error during the operation of the plant. There have been thousands of accidents in nuclear power plants during the past 40 years.

The worst accident so far occurred at Chernobyl in Ukraine, where a steam explosion and partial core meltdown resulted in radioactive contamination of the entire Northern Hemisphere. The effects of this disaster include increased cancer rates and genetic damage, whose full effects will not be known for many generations.

Equally serious are the problems of old and outdated nuclear plants and the highly radioactive waste created by both commercial and military reactors. No one has found an acceptable solution to the problem of dismantling shutdown reactors and their containment buildings. They stand as festering mausoleums that are structurally unsound and pose a long-term threat to the communities in which they are located.

Over the years, a mountain of high-level radioactive waste has accumulated at a large number of sites across the country. The metal drums in which the waste is stored are rusting and leaking, contaminating soil, groundwater, and nearby waterways. The problem gets significantly worse each year. In 2002, the federal government certified Yucca Mountain, Nevada, as an acceptable burial site for these wastes, a decision being contested in court by the State of Nevada. However, whichever site is eventually chosen, no one can guarantee safety for 10,000 years, the estimated minimum time the wastes must be kept from contact with plant and animal life. No human civilization has lasted for 10,000 years.

# 10
## Conclusion

I am only one; but still I am one. I cannot do everything, but still I can do something. I will not refuse to do the something I can do.
—Edward E. Hale

So where does this leave us? What should we conclude from all this? Are America's environmental problems already insurmountable, or can we change course and solve them? Do we know how? Do we want to?

### The Public's View

The time is always right to do what is right.
—Martin Luther King, Jr.

Americans clearly want something done to restore the relatively clean environment we inherited from our ancestors (figure 10.1, table 10.1). They are quite conscious of air and water pollution and are willing to pay to stop them. They know the nation has been and still is committing environmental suicide and want something done about it. No one plans to live in a cold dark cave, but most believe that comfort, a good life, and environmental protection are not either-or choices.

### Industry's View

A human [is] not a fallen god, but a promoted reptile.
—J. Howard Moore, *The Universal Kinship*

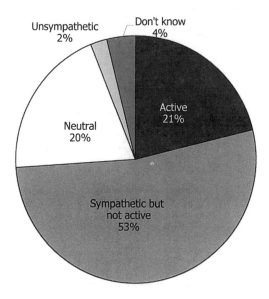

**Figure 10.1**
How Americans rate themselves as environmentalists (National Environmental Education and Training Foundation).

America has a very productive capitalist economic system. Few Americans are interested in converting to socialism. Capitalism, combined with representative government, has produced the highest standard of living the world has ever seen. Our affluence is the envy of the world. The question we face is whether businesses that must always keep at least one eye on profitability and their stockholders can afford to spend money on environmental concerns. Is this "wasted" money that adversely affects the "bottom line"? Will it cost jobs? Is the phrase "responsible capitalism" an oxymoron?

Americans share three common beliefs about environmental regulations and the economy. One is that environmental rules cause widespread unemployment. Another is that environmental regulation has led to many plant shutdowns and aggravated unemployment at the local level. And a third belief is that environmental regulation has caused lots of companies to build new plants overseas where they can escape onerous environmental rules. However, when unemployment and economic data from the past 30 years are analyzed, all three beliefs turn out to be false. Nearly all economists agree that these three beliefs have no basis in reality.

**Table 10.1**
What the American public believes about existing laws that deal with
environmental problems

| | Type of environmental problem | | | | |
|---|---|---|---|---|---|
| | Water pollution (%) | Air pollution (%) | Protecting wild or natural areas (%) | Protecting wetlands (%) | Protecting endangered species of plants and animals (%) |
| Have not gone far enough | 72 | 62 | 48 | 44 | 41 |
| Have struck about the right balance | 19 | 24 | 32 | 27 | 33 |
| Have gone too far | 4 | 8 | 13 | 15 | 21 |
| Don't know | 6 | 6 | 7 | 14 | 5 |

*Source:* National Environmental Education and Training Foundation.

A good example of the attitude of industry through most of the twentieth century is provided by the response of the automobile industry in the late 1960s to the Clean Air Act, passed over their strong opposition in 1970. The industry has an important role in our economy, so any threat to it is taken seriously by legislators, who are rightfully fearful of damaging economic growth. Auto executives used this fear to manipulate policymakers. The Automobile Manufacturers Association said that "to achieve the control levels specified in the bill . . . [M]anufacturers . . . would be forced to shut down." Lee Iacocca, as vice president of the Ford Motor Company, warned that, should the Clean Air Act become law, "We could be just around the corner from a complete shutdown of the U.S. auto industry . . . " He called the Clean Air Act "a threat to the entire American economy and to every person in America." He added that the emission standards were "in two words: IM POSSIBLE." As a final comment, he said "We've got to pause and ask ourselves: How much clean air do we need?"[1]

Despite Iacocca's prognostications, which, of course, were mistaken, the Clean Air Act, now recognized as an essential cog in the wheel of environmental protection and cleanup, became law. However, the automakers subsequently convinced Congress and the EPA that limits could not be met and the automakers were granted extensions. These extensions were granted to American automakers at the same time foreign manufacturers were meeting and exceeding the EPA's requirements.

Many American companies are starting to realize the falsity of their assumptions about the effect of environmental rules on their financial health. They are beginning to build environmental considerations into their business plans. They spend over $1 billion a year on out-of-house media advertisements, and those that deal with the environment aim to convince the public of their leadership in the area of environmental sensitivity. There is more recognition by companies that there may be an economic advantage to reducing emissions of greenhouse gases.

**Who Is to Blame?**

We're all in a great big car driving at a brick wall at 100 mph and everybody is arguing over where they want to sit. My point is it doesn't matter who's driving. Somebody has got to say, "For God's sake, put the brakes on and turn the wheel."
—David Suzuki, ecologist

Answering this question is like answering the same question about cocaine addition. Is it the producers in Colombia, Myanmar, Afghanistan, and elsewhere? Or is it the consumers, mainly American, who demand that their habit be satisfied? Which is the chicken and which is the egg, and does it really make any difference? Clearly, both groups must share the guilt and both groups must decrease their harmful activities.

**Water Supplies**
There is no question that Americans pay little attention to the amount of water they use. It is so cheap in the United States that it is of little concern to most people outside of the decreasing number of grain-growing farmers and those living in a few dryland cities. For most Americans water shortages are only something they hear about in TV stories dealing with northern Africa and the Middle East, and water pollution is a problem centered in Haiti or perhaps somewhere in Southeast Asia. Americans tend to dismiss such problems as concerning only the UN or the World Bank.

Unfortunately, this is not the case. Many of America's cities are on the verge of serious water shortages and underground aquifers are increasingly depleted. One day in the near future newspaper and TV stories will be emphasizing water problems at home rather than those abroad. Part of the political problem in looking for solutions is that democracies are ambulance forms of government, not preventive-medicine forms of government. Nothing is done until the patient is near death, when panic sets in. The patient is not yet near enough to death for most Americans to become unnerved. This point of view is hardly surprising. Do you visit the doctor when you are well or have only a bad cold? Probably not. Your body can likely cure itself. Unfortunately our water problems will not cure themselves.

But we can do something as individuals in our own homes, as outlined in chapter 1. The changes suggested are not onerous and are easily implemented. The only things that cost money are new faucet washers, a low-flow showerhead, and some low-flow toilet tanks if you don't already have them (ask a plumber). Over a brief time these changes will pay for themselves, and after that it's all profit. Concern is nice and talk is cheap. Only action will help. Do it!

In addition, we can ask our local and state governments about leaky water pipes. I hate taxes as much as anyone, but allowing water mains to pour water into the ground instead of into our homes is like flushing clean water down the drain. Why do it? The cost of laying new pipe is an investment, not an expense. It will start paying dividends immediately for you, for your children, and for their children for many generations.

Ask your city authorities about water pricing, sewage treatment, and water recycling. There may well be things you would like to see changed. Lower prices in the winter and higher prices in the summer? Higher rates for profligate users of an increasingly scarce resource? Repair or upgrading of an aging water-treatment plant? Do you care about your community?

## Floods

Is your city located near a river that floods "every now and then?" Do you, or anyone in city government, know what the 50-year or 100-year floodplain is? Shouldn't someone? Is your property in danger? Has the state geological survey (usually located in the state capital) prepared a flood-hazard map for your community? Some states have such maps for potentially endangered areas, but most states don't. Is this something you would like to see done in your state?

What about zoning laws in floodplain areas? Would you like to see such laws enacted in your state? Can they be made fair to those already located on the floodplain and to real estate interests that want to develop the floodplain? Does your city have a detailed emergency plan that is widely circulated among the citizenry? Should it? Let your congressional representatives know your views on subsidizing rebuilding in flood-prone areas.

### Garbage

Does your community have a landfill? Is it near capacity yet? Does it conform to the latest safety standards? Could methane be recovered from the landfill? If your garbage is incinerated, what controls are in place to prevent noxious fumes and smoke from polluting the air? Do you separate hazardous materials you discard from the rest of the trash? Do you know what to do with it when you do? Does the city government care about such things? If not, would you like it to? What does your city pay for hauling the garbage away from residential areas? Has there been discussion of the costs and benefits of recycling in your community?

Is there a polluted site in your area that has been designated as a Superfund site? It doesn't have to be in your neighbor's backyard to affect you. Do you or does your community get any water from wells? Water moves and, depending on the character of the aquifer, can move many feet per year. Are your children drinking water laced with arsenic or lead? What office in the city analyzes the water you drink? How frequently is the water tested?

### Soil and Food

Farmers are normally well aware of the costs of production and the problem of soil erosion on their cropland. It is a simple matter of survival for them. If they don't know about such things they will not be in business long. But have you, if you have a garden, considered what those pesticide sprays you use are adding to your soil and plants? Have you considered an organic garden? Wouldn't it make you feel better, both psychologically and physiologically, to stop eating pesticide chemicals that have become part of the fruit, chemicals that cannot be washed off?

Do you have strong feelings about genetically modified foods? Let your congressional representatives know your views. Should the altered foods be labeled "GM," or is such labeling simply an unnecessary burden on an already-stressed industry? Those in Congress are sensitive to the views of their constituents on such contentious issues. They want to be reelected. Let them know what you think.

### Energy Sources

It is true that the United States, with less than 5 percent of the world's population, consumes 25 percent of the earth's energy resources, mostly fossil fuels. The question often asked is: Is this fair to the other 95 percent of the human population? Offsetting most people's normal response of "no" is the realization that the United States probably generates at least 25 percent of the world's economic activity. If our energy sources were changed from fossil fuels to solar power or wind power, which are inexhaustible, there would be no complaints by other nations about usage. For a nation that wants to be liked, not just envied, this is the way to go. Such a change is the only realistic way to end the growing hostility toward America because of its overuse of the world's energy resources.

The nineteenth century was the century of coal. The twentieth belonged to oil. But as the twenty-first century dawns, it is recognized worldwide that the burning of coal and oil needs to be seriously curtailed. These fossil fuels are the cause of most climate change and water- and air-pollution problems. But such a change will be time consuming, difficult, and expensive, and will be strenuously opposed by large and well-established industries, a certain recipe for federal foot-dragging.

The "old" nonpolluting renewables, hydropower and geothermal power, are geographically and geologically limited and no one believes they can form the major part of America's energy base. Biomass is renewable and, if only dead vegetation is used, does not add to the enhanced greenhouse effect. (Natural decay would release the same amount of carbon dioxide gas.) But biomass cannot yield enough energy to power American industry. Hope at the moment for clean energy is based on wind power, solar power, and hydrogen power, sources that are not independent. Both wind power and solar power can be fed into an electrical grid and used to undergird hydrogen power. Hydrogen only exists on the earth as part of chemical compounds, and energy is needed to free it.

President Bush's fiscal-year budget for 2004 gave top priority to coal and nuclear activities, while funding for most energy-efficiency and renewable–energy programs remained flat or even declined.[2] The one exception was an ambitious new plan to develop hydrogen-fueled vehicles and the necessary infrastructure to produce, store, and distribute hydrogen fuel. Unfortunately, the president's plan emphasized fossil fuel and nuclear sources to produce the hydrogen. Wind power was slated to receive a 5.5 percent decrease, biomass/biofuels fell 19 percent, geothermal technology dropped 3.8 percent, and solar energy received a negligible 0.1 percent increase. The energy portion of the fiscal-year 2004 federal budget was a particularly clear example of the triumph of power politics over the expressed wishes of the American people. It's enough to bring tears to a citizen's eyes, unless that citizen is an executive of an oil company.

Lest we heap too much scorn on Republicans, Democrats are not without faults in the alternative-energy debate. For example, wind energy is particularly strong off the Massachusetts coast (figure 5.7), and Cape Wind Associates has proposed erecting a wind farm consisting of 130 turbines that will rise 426 feet above the water 6 miles off the coast of Cape Cod. At that distance the turbines will be visible from beaches along Cape Cod, Martha's Vineyard, and Nantucket, where some of the nation's wealthiest liberals own valuable property. The wind farm would provide three-quarters of the Cape's electricity, pollution-free. The Kennedy family, normally strong proponents of alternative energy but owners of a major compound in the area, has been leading the charge against the proposed wind farm. They have equated it with erecting the turbines in the Grand Canyon. NIMBY is a disease that apparently can affect even those who push most strongly for alternative sources of energy. The rubber is hitting the road at Hyannis and causing much screeching.

Only a concerted and widespread public outcry can increase the glacial speed at which pollution-free energy sources are presently being developed in the United States. Massive federal subsidies are needed, a need that impacts the federal budget. If tax increases are to be avoided, and we all favor that, existing subsidies to the fossil-fuel industries need to be slashed or eliminated. When reelection time nears, your representatives may rely on the cash they get from the fossil-fuel lobby to finance their campaigns, but they also need your vote. It is a cliché to say "write your congressperson," but without public pressure the rapid development of wind, solar, and pollution-free hydrogen energy is unlikely to happen within your lifetime.

The Sustainable Energy Coalition has recently devised a simple set of questions about America's use of energy for your executive and legislative representatives or candidates to answer if they want your vote (appendix A). Why not send it to them?

## Global Warming

Most of the things an individual can do to slow global warming revolve around car usage. But there are several problems with galvanizing citizen action on this front. First, because only a small percentage of exhaust pipes emit visible smoke, we are rarely conscious of the stuff that comes out of the pipe whenever the motor is running. The exhaust gases from most cars are invisible. Out of sight, out of mind. Second, except for those who live permanently near the seashore from Texas to Virginia, sea-level rise and the frequency of hurricanes are not major concerns. Third, and most significant, Americans tend to view their cars as extensions of themselves. An invisible plaque that reads "KEEP YOUR GREASY HANDS OFF MY CAR" is present in most American homes. Thanks in large part to advertising, cars represent status, sex, and other unrelated things rather than simply a fast way to get somewhere. The most disastrous thing that could happen to the auto industry would be for Americans to realize that the essentials of a car are the motor, transmission, wheels, and safety features rather than deliberately ephemeral styles, a representation of your place in society, or an indication of your worth as an individual. A small car will get you anywhere as fast as a large car, given the speed limits on American roads and the ever-worsening traffic congestion in American cities. In 2000, New York became the first state where residents traveled for more than 30 minutes *on average* each way between their jobs and homes. In Los Angeles, rush-hour speed averages 32 mph and the average driver in 2001 spent 90 hours, nearly four days, trapped in rush-hour traffic. The time to travel between any two points has doubled in the past 2 decades. Other blighted areas included San Francisco–Oakland at 68 hours; Denver, 64; Miami, 63; and Chicago and Phoenix at 61.

On September 22, 2003, tens of millions of Europeans got a brief respite from noxious car fumes as the sixth annual car-free day was celebrated (or tolerated) in nearly 1,000 European cities. Permitted transportation choices on this day are buses, bicycles, and walking. Emergency vehicles are exempted.

In 2003, Ken Livingstone, the mayor of London, put his political career on the line by reclaiming his city from the tyranny of the automobile.[3] He drew a lopsided oval line around the city and announced that almost anyone who moved or even parked a car on the street within it after February 17 would be fined 5£ (more than $8) a day for every day it happened. If a driver failed to pay, one of more than 700 vulturelike cameras (originally developed for antiterrorism purposes in London) perched throughout the zone would capture the person's license-plate number and relay it to a computer, leading to a $180 fine. The first day of the charge, the number of cars within the exclusion zone dropped by about 60,000 (a number that has remained stable), from 250,000 to 190,000, and average speed in central London doubled. London's bookies set odds of 10 to 1 that the mayor would be thrown out when his term expires. They were wrong.

In New York, more than 250,000 vehicles crowd into the 8.5-square-mile heart of Manhattan in 3 hours every morning, roughly the same number that enter the 8 square miles of central London in an entire workday. Will Mayor Bloomberg pursue the Livingstone Initiative?

Another change you might make involves your choice of electric utility. Deregulation of the utility industry is taking hold, and your state may allow you to choose a supplier that uses an energy source other than coal to generate electricity. Thirty states have revised their laws to promote various types of clean energy. Pennsylvania and California let customers choose their power source and, among those who have switched, 20 percent in Pennsylvania have selected green power. In California, 95 percent chose green sources. It probably will cost you a few dollars more each month, but it seems a small price to pay to contribute to slowing global warming and, perhaps more importantly, lowering air pollution.

## Air Pollution

Anything that stops noxious gases from polluting our air is beneficial. This includes buying smaller cars (Ha! Fat chance!) and keeping them tuned. Getting electricity from a supplier who doesn't use coal will definitely have an effect as well if enough people do it. The solution to the air-pollution problem is similar to the solution to the climate-change problem. Stop using fossil fuels. Unfortunately, the current federal government believes otherwise. In August 2003, bowing to intense pressure from industry, they decided to allow thousands of older power plants, oil refineries, and industrial units to make extensive upgrades without having to install new antipollution devices. Several states have challenged this ruling in court.

At present, wind and solar power combined contribute less than 1 percent of America's energy mix (hydrogen power is 0) and, barring a sharp change in course by the federal government, this percentage will grow only very slowly.

## Ozone

Although the production of ozone-destroying chemicals has ended in the United States and will soon end everywhere, the ozone hole will not disappear completely for about 50 years. Keep this in mind during summer months. Wear a wide-brimmed hat and use a sunscreen lotion, especially when you are seriously underclothed, as at the beach. Some women's cosmetics now contain sunscreen protection. Check the label. Skin cancer was a significant problem before thinning of the ozone layer and is even more significant today. The effect of the sun's ultraviolet rays on your skin is similar in one respect to smoking cigarettes. You are unlikely to be affected before age 50, which for most young people is unimaginably far away. Youth feels immortal.

## Nuclear-Waste Disposal

There's not much you can do about this problem, other than switching from a nuclear electricity supplier to one that uses another source, assuming your state gives you a choice. If enough folks do this, the plant may be forced to close down. You can also find out whether any radioactive waste is stored in your vicinity. Ask the city government or your state's geological survey agency. Every state has one. You can also find out whether there are regular inspections for leakage from storage drums around the nuclear plant. If there are, what did they find, and if storage canisters are even rusty, what is being done about it? If you are unhappy with the answers to your questions, screams, letters, and e-mails to your congressional representatives are in order. It is not a good idea to take chances with radioactivity.

## Other Environmental Problems

Population increase, urbanization, technology, and poverty in most of the world have created, and will continue to create, innumerable environmental problems. New environmental concerns will, however, be of minor importance compared with the ongoing pollution of our water, air, and soil. Both print and visual media will bring new problems to our attention because this helps circulation and sells advertising space.

Unfortunately, it is not always easy for a concerned citizen to get a balanced and accurate treatment of environmental problems. Journalists and the authors of books are normal human beings. They differ from other people only in having a greater facility with words. Journalists often do not have the time to investigate an issue thoroughly (Hurry, the presses are rolling!) and even when they do, may make mistakes in interpreting what they find out. And, sad to say, the reporter may have an ax to grind and may be less than impartial in telling a story. Usually this results from having a political point of view and may not be a conscious or deliberate attempt at bias.

## Governmental Policy in the United States

The men who cared so much for the future, who were so concerned about the establishment of rights against infringement by government or individual, these visionary men *forgot* to establish your right to breathe clean air or drink potable water.
—Victor J. Yannacone

America is governed by elected officials and, in principle, our wishes should quickly be translated into effective legislation. If the public wants additional controls on pollution of our air and water, they should get them. Unfortunately, this is not always what happens because of the influence exerted on legislators by what are commonly called "vested interests." These are individuals, special interest groups, and corporations who have money to spend, believe they are benefiting from the way things operate at present, and fear they will be damaged by proposed changes.

There is nothing wrong with people or organizations trying to protect what they see as their best interests. The problem is that not everyone has equal access to elected officials. Clearly, if you have money you are more likely to get your way than if you are penniless. A person or organization that donates a million dollars to a political campaign can be sure of speaking directly to a candidate when he or she phones. Can you? As they say, money talks, and it influences not only legislators but the general public as well.

An obvious example of the ability to influence the general public is the ability to advertise. Industries with strong economic interests may emphasize the views of the minority of scientists who doubt the reality of the human influence on global climate change. Alternatively, they may emphasize the opinions of those who believe that environmental concerns are being overblown by a small group of "econuts," "tree huggers," or political leftists who want governmental control over every aspect of our lives. Scenes of happy people romping in beautiful scenery that has not been damaged by an oil-drilling platform nearby may convince people that petroleum is benign. Did the producers of Marlboro cigarettes advertise that the famous rugged-looking, cigarette-smoking Marlboro Man of a few decades ago died of lung cancer? If they did I must have missed it.

### Environmental Taxes

Governments tax income because it is an easy way to raise revenue, not because it serves any particular social goal other than redistributing income. Taxing environmentally destructive activities, however, both raises revenue and is socially constructive. There is a broad body of research demonstrating that environmental taxes change fuel choices. A 1998 public opinion poll revealed that 71 percent of American voters favor tax shifting as a way to reform the tax structure. They support a shift from taxing work (payroll taxes) and production (corporate taxes) to taxing

**Table 10.2**
Gasoline tax rate per gallon, 2003

| State | Tax (cents) | Rank |
|---|---|---|
| Alabama | 18.0 | 36 |
| Alaska | 8.0 | 49 |
| Arizona | 18.0 | 37 |
| Arkansas | 21.5 | 22 |
| California | 18.0 | 38 |
| Colorado | 22.0 | 18 |
| Connecticut | 25.0 | 8 |
| District of Columbia | 20.0 | |
| Delaware | 23.0 | 14 |
| Florida | 14.1 | 47 |
| Georgia | 7.5 | 50 |
| Hawaii | 16.0 | 43 |
| Idaho | 26.0 | 4 |
| Illinois | 19.8 | 31 |
| Indiana | 15.0 | 45 |
| Iowa | 21.1 | 26 |
| Kansas | 23.0 | 15 |
| Kentucky | 16.4 | 42 |
| Louisiana | 20.0 | 27 |
| Maine | 22.0 | 19 |
| Maryland | 23.5 | 13 |
| Massachusetts | 21.0 | 24 |
| Michigan | 19.0 | 33 |
| Minnesota | 20.0 | 28 |
| Mississippi | 18.4 | 35 |
| Missouri | 17.0 | 40 |
| Montana | 27.0 | 3 |
| Nebraska | 25.5 | 6 |
| Nevada | 24.0 | 10 |
| New Hampshire | 19.5 | 32 |
| New Jersey | 14.5 | 46 |
| New Mexico | 18.9 | 34 |
| New York | 22.6 | 17 |
| North Carolina | 23.6 | 12 |
| North Dakota | 21.0 | 25 |
| Ohio | 22.0 | 20 |
| Oklahoma | 17.0 | 41 |
| Oregon | 24.0 | 11 |

Table 10.2
(continued)

| State | Tax (cents) | Rank |
|---|---|---|
| Pennsylvania | 25.9 | 5 |
| Rhode Island | 31.0 | 1 |
| South Carolina | 16.0 | 44 |
| South Dakota | 22.0 | 21 |
| Tennessee | 21.4 | 23 |
| Texas | 20.0 | 29 |
| Utah | 24.5 | 9 |
| Vermont | 20.0 | 30 |
| Virginia | 17.5 | 39 |
| Washington | 25.4 | 7 |
| West Virginia | 23.0 | 16 |
| Wisconsin | 28.1 | 2 |
| Wyoming | 14.0 | 48 |
| AVERAGE | 20.4 | |

*Source:* Tax Policy Center.

pollution and consumption. Most economists agree, but this would be a monumental change in tax policy by the federal government. Green economists would purge the tax code of regulations and loopholes that clearly encourage environmental degradation. New levies would be applied on pollution generators like products containing lead, gas-guzzling cars, ozone-depleting chemicals, and the burning of coal and oil. But the public must be convinced that the changes are really only a shift of the tax burden, not a subterfuge for an overall tax increase. This will not be an easy sell, given the record of past governments.

America has had environmental taxes for a long time. We simply have not thought of them as such, only as revenue sources for local, state, and federal governments. The clearest and most obvious example is the tax on gasoline. In 2000, it averaged 41 cents per gallon, the state taxes averaging 20.4 cents and the feds getting 18.4 cents (table 10.2). This state and federal tax burden on gasoline, adjusted for inflation, has not changed significantly since 1991.[5] As shown in table 5.5, our low tax on gasoline is the envy of many car owners in the Western world, but not of those with an environmental conscience. Some states have a whole raft of taxes aimed not only on fossil-fuel use but at air pollution, solid-waste disposal, water pollution, and other environmentally significant concerns. Massachusetts has 14 environmental taxes, fees, or exemptions.[6] California and Minnesota are two other states where green taxes are taking root.

**The Carbon Tax**    A recent example of environmental legislation that is badly needed in the United States but did not win congressional approval when proposed in 1997 is the carbon tax. The concept behind the proposed tax is that the only way to get people to reduce their use of energy (overwhelmingly coal and oil) is to make energy more expensive. The wallet is very sensitive to thickness. The more carbon-based energy fuel you use (coal, oil, natural gas), the more tax you pay. The burden of this tax would fall initially and most heavily on coal-burning utilities and industrial users and processors of petroleum in its various forms. It would be no fun for car owners either.

Naturally, this proposed tax was strongly opposed by industry, with some support from segments of the general public. Their fight was successful. The tax did not make it through Congress. However, the sooner we all understand that there is no such thing as a "win-win" solution to carbon (dioxide) emissions, the better. There will be an initial cost involved in reducing emissions, but there is no question that the reduction will pay for itself many times over in human health. Do you like cancer? I don't. Do you like dirty lungs? I don't. Do you like brain damage in your children? I don't.

The view that repayment for the carbon tax will be forthcoming is not a "pie in the sky" hope put forth by people who like taxes because it gives government more control over our lives. In economic terminology, pollution control is an investment, not an expense. The money spent is certain to bring financial dividends, most clearly in the cost of maintaining human health. Are you willing to make this investment? What is your health and that of your children worth?

The argument that environmental taxes will cost jobs and lower America's competitiveness is simply not supported by data accumulated by those who have studied the question. A 1997 study by the World Resources Institute found little evidence supporting the view that environmental taxes, "green taxes," or "ecotaxes" harm the economy. An economist at the Center for Economic Studies reported that "there is simply no evidence that superior environmental performance puts firms at a market disadvantage."

Nevertheless, those opposed to changes in tax law continually trumpet the possibility of damage to our capitalist economy by taxes friendly to the environment. Interestingly, the American public seems not to believe them. Nearly two-thirds of us (63 percent) in a 2000 poll believe that environmental protection and economic development can go hand in hand. And when asked to choose between the two, 71 percent favor the environment.

Though most politicians pay lip service to the environment, getting a sweeping tax shift through Congress is no easy task. Former Treasury Secretary Robert Reich likes the concept of green taxes but doesn't see it becoming policy soon. "I wish I could be optimistic, but politically, it's a very hard sell," he says. "Energy states are very powerful in Congress." They will not quietly accept the phasing out of environmentally damaging subsidies such as those for fossil fuels.

Nevertheless, there is reason to hope that there will eventually be a shift toward green taxes and away from taxes on payroll, personal, and corporate income. These sources currently supply 90 percent of federal revenue. As put by the Worldwatch Institute, "For Progressives [liberals], [tax shifting] has the appeal of protecting the environment by making the polluter pay and reducing unemployment. For conservatives, it offers the advantage of using the market, rather than regulatory agencies, to protect the environment, and allows for cuts in much-resented income or sales taxes that may inhibit constructive economic activity."

### Green Taxes in Other Countries

Think globally, but act locally.
—René Dubos, *Celebrations of Life*

Taxes on environmentally harmful activities are levied widely in Europe and total 7 percent of total tax revenue.[7] Eight countries have instituted carbon taxes, nine tax waste disposal, and others levy green taxes on batteries, packaging, and car tires. However, nearly all environmental tax revenue comes from just two sectors, energy and transport.

Many European countries levy a charge on water pollution. In the Netherlands, a surface-water-pollution tax has been levied since 1969, based on the level and toxicity of discharges. The revenues are used to build water-treatment plants. The rate of the tax is determined by the cost of the measures to counter and prevent pollution of surface waters. Introducing this tax has resulted in large polluters making cost savings by cleaning their own wastewater before discharge. The Netherlands also has energy and environmental taxes on fuels.

In Sweden environmental taxes have been gradually phased in since 1984. They include taxes on energy consumption, carbon dioxide, sulfur dioxide and nitrous oxide emissions, fertilizer, pesticides, the scrapping of cars, and water pollution. The Swedes have been careful to recycle the revenues back into subsides for green investments. Revenue from the taxes has not simply been used to swell government coffers.

Interestingly, the revenue starts coming in even before the tax takes effect. In 1992 Sweden instituted a tax on emissions of nitrous oxide on power stations, accompanied by regulations requiring installation of modern combustion technology. Reductions in nitrogen emissions began immediately after the Swedish government announced its plans in 1990. By the time the tax was instituted in 1992 nitrogen emissions had already dropped by 35 percent.

Denmark has been ratcheting up its carbon dioxide tax since 1996. Tax revenues are recycled into support for energy efficiency and rebates on labor taxes. Heavy industrial energy users can obtain partial rebates by entering into energy-efficiency agreements with the Danish government. Denmark has also instituted a waste tax and a tax on sulfur in fuels.

A framework for providing ecotaxes was introduced in Belgium in 1993 and was designed to change the purchasing patterns of consumers in favor of more environmentally friendly alternatives. The level of the tax aims to reflect environmental damage but is high enough to put pressure on producers and consumers to alter their behavior. Taxation applies to all drink containers; disposable razors and cameras; industrial packaging; batteries; pesticides and plant-protection products; and paper. As in Sweden, the Belgian government found that announcing future levies provokes immediate action by industry, even before the tax takes effect.

In 1996 Britain imposed a landfill tax of £7 per ton for decomposable waste and £2 per ton for inert waste. In 1999 the tax rate for decomposable waste was increased to £10 per ton, and this will rise by £1 per ton each year to £15 per ton in 2004, when it will be reviewed again. Operators have the option of making their payments to environmental trusts. If they do this they can claim rebates on their landfill tax bills of 90 percent of their contributions to the trusts, up to a maximum of 20 percent of the tax bill. Revenue from the landfill tax is used to reduce payroll taxes. Six other countries in Europe have adopted waste taxes as well. In Denmark the amount of waste brought to landfills and incinerators has fallen 26 percent and recycling is up significantly.

In 1999 Britain imposed a new tax on energy used by industry. A new tax was also levied on fuel used for transport, and incentives were included to encourage people to drive smaller cars. Most British cars are already small compared to those driven by Americans.

France has carbon dioxide and sulfur dioxide taxes and a landfill tax, with the funds flowing into environmental investments. Both France and Germany tax water pollution and use the revenues to build new and better waste-treatment plants. Germany has a tax on items as varied as hazardous wastes and disposable fast-food packaging. In 1999 Germany passed a carbon tax similar to the one defeated in the U.S. Congress a few years ago. The tax was made financially neutral through reductions in social security contributions. Italy has operated a carbon/energy tax since 1999.

Fifteen European countries have adopted green tax approaches, as have nations in Asia, including South Korea, Taiwan, and Singapore. In regard to ecotaxes, these nations are more progressive than the United States. They are reducing taxes on wages and personal income and replacing them with taxes on things that harm the environment. So far, the amount of tax shifting has been small, generally less than 2 percent. But it is a start in the right direction.

Summarizing, it is clear that nearly all of our environmental problems are political and social problems, not scientific problems. Scientists know how to clean up most environmental problems, although some uncertainties remain. There are two barriers to accomplishing it. The first is the seeming reluctance of Americans to insist that their federal and state representatives pass and enforce strict antipollution laws. We can ask our representatives how they voted on specific environmental legislation and why. Do they object to shifting the tax burden from our (and their) personal income taxes to taxes on activities that harm our environment?

If they do not object, perhaps you can form a local committee to work with them on this type of reform. The Internet contains a wealth of information about ecotaxes and their effects in states and countries where they have been instituted.

The second barrier to environmental remediation is public apathy in regard to such things as reducing usage of water, recycling waste, buying smaller cars, and switching to alternative sources of energy. If we act we can accomplish the changes that we tell the pollsters we want. In brief, we have caused the environmental problems we now face but we can solve them. We can make the necessary lifestyle changes. They really are not onerous and will greatly increase the quality of life for us all.

Thomas L. Friedman, syndicated columnist for the *New York Times*, wrote a column that appeared shortly after the tragedy of September 11, 2001, titled "Ask Not What. . . ." He was referring to the famous inaugural address by President John F. Kennedy on January 20, 1961, that sought to inspire the country to sacrifice for the greater good. Friedman was imagining what would have happened if President Bush had challenged the country to conserve energy, improve mileage standards, and sacrifice for national goals. He felt that the nation's second "greatest generation" may be in the wings just waiting for such a chance to prove their mettle. Let us hope he is right.

Never doubt that a small group of thoughtful, committed citizens can change the world. Indeed, it's the only thing that ever has.
—Margaret Mead, anthropologist

I can't understand why people are frightened of new ideas. I'm frightened of the old ones.
—John Cage

# Appendix A
## Sustainable Energy Coalition

Presidential Candidates Survey on National Energy Policy Issues

**I. Energy Efficiency/Renewable Energy – Budget and Tax Policy**

1. Do you support increasing, maintaining, or reducing the funding levels for the U.S. Department of Energy's energy efficiency and renewable energy programs (likely to be funded at approximately $1 billion in fiscal Year 2004)?

_____ INCREASE

_____ REDUCE

_____ MAINTAIN

Comment:


2. Do you support federal tax incentives to encourage consumers to purchase products such as cars, homes, appliances, and heating and cooling systems that are very energy efficient but which might otherwise be more expensive?

_____ YES

_____ NO

Comment:


3. Do you support federal tax incentives to expand the use of hybrid cars in this country?

_____ YES

_____ NO

Comment:

4. Do you support continuation of the federal tax incentives for ethanol?

_____ YES

_____ NO

Comment:


5. Do you support increasing, decreasing, or maintaining the current level of federal tax incentives and/or federal budget outlays to promote the use of solar energy technologies?

_____ INCREASE

_____ DECREASE

_____ MAINTAIN

Comment:


6. Do you support increasing, decreasing, or maintaining the current level of federal tax incentives and/or federal budget outlays to promote the use of wind energy technologies?

_____ INCREASE

_____ DECREASE

_____ MAINTAIN

Comment:


7. Do you support increasing, decreasing, or maintaining the current level of federal tax incentives and/or federal budget outlays to promote the use of biomass/biofuels energy technologies?

_____ INCREASE

_____ DECREASE

_____ MAINTAIN

Comment:


8. Do you support increasing, decreasing, or maintaining the current level of federal tax incentives and/or federal budget outlays to promote the use of geothermal energy?

_____ INCREASE

_____ DECREASE

_____ MAINTAIN

Comment:

9. Do you support increasing, decreasing, or maintaining the current level of federal tax incentives and/or federal budget outlays to promote the use of hydropower energy technologies?

_____ INCREASE

_____ DECREASE

_____ MAINTAIN

Comment:


10. As the federal government expands its hydrogen development program, questions have been raised as to which fuel sources should be relied upon to produce the hydrogen. As a percentage of funding, how should the federal government allocate its resources among the potential hydrogen production sources?

_____ % RENEWABLES (i.e., solar, wind, geothermal, biomass, hydropower)

_____ % FOSSIL FUELS (i.e., oil, coal, natural gas)

_____ % NUCLEAR POWER

Comment:


**II.   Energy Efficiency/Renewable Energy – Regulatory Policy**

1. Do you support, in federal energy legislation, a Renewable Portfolio Standard (aka renewable energy standard) that would require the federal government and the states to ensure that electricity generators provide a portion of their power from renewable energy sources such as solar, wind, geothermal, incremental hydropower, and biomass (e.g., 20% by 2020)?

_____ YES

_____ NO

Comment:


2. Do you support, in federal energy legislation, a Renewable Fuels Standard for transportation fuels that would require the federal government and states to ensure that a percentage of transportation fuels is provided by renewable energy sources such as ethanol (e.g., 5% by 2010)?

_____ YES

_____ NO

Comment:

3. Do you support, in federal energy legislation, a small wire charge on everyone's electric bill (e.g., 2–3%), as previously included in most utility rate bases, to maintain funding of energy efficiency, low-income weatherization and energy assistance, and renewable energy research, development, demonstration, and deployment programs?

_____ YES

_____ NO

Comment:


4. Do you support increasing, decreasing, or maintaining the current level of federal purchases of green electricity and/or decentralized renewable energy technologies and energy efficiency measures?

_____ INCREASE

_____ DECREASE

_____ MAINTAIN

Comment:


5. Do you support mandatory federal policies to enable distributed generation technologies such as fuel cells and renewable energy to connect to the electricity grid?

_____ YES

_____ NO

Comment:


6. Do you support raising the Corporate Average Fuel Economy (CAFE) standards for new cars, Sport Utility Vehicles (SUVs), and other light trucks to a combined fleet average of at least 40 mpg by the year 2010?

_____ YES

_____ NO

Comment:


**III.   Climate Change**

1. Do you believe the current level of scientific evidence that human activity (i.e., the combustion of fossil fuels and the production of greenhouse gases) is causing global warming and warrants immediate precautionary actions?

_____ YES
_____ NO
Comment:


2. Do you believe the United States has shown adequate international leadership on global warming?
_____ YES
_____ NO
Comment:


3. Do you support ratification of the Kyoto Protocol to curb greenhouse gas emissions?
_____ YES
_____ NO
Comment:


4. What binding $CO_2$ emissions reduction measures, if any, will you support in your Presidency?
Comment:


## IV.  Fossil Fuels
1. Do you support increasing, reducing, or maintaining the current level of tax incentives available to the domestic coal, oil, and natural gas industries?
_____ INCREASING
_____ DECREASING
_____ MAINTAINING
Comment:


2. Do you support permanently protecting the Arctic National Wildlife Refuge (ANWR) from oil exploration and drilling?
_____ YES
_____ NO
Comment:

3. Do you support federal regulation of $CO_2$ emissions by fossil-fueled power plants and/or other sources?

\_\_\_\_\_ YES

\_\_\_\_\_ NO

Comment:


4. The U.S. Senate passed a bill committing the nation to reduce oil consumption by 1 million barrels per day by 2013. If you support this legislation, what actions will you take by the end of your first term to achieve these, or additional, savings?

Comment:


5. What actions would you take during your first term to address the projected natural gas supply shortage and price increases?

Comment:


V.   Nuclear Power

1. Do you support the construction of new nuclear power plants?

\_\_\_\_\_ YES

\_\_\_\_\_ NO

Comment:


2. Do you support the relicensing of existing nuclear power plants in the United States?

\_\_\_\_\_ YES

\_\_\_\_\_ NO

Comment:


3. Do you support the U.S. Senate proposal to extend federal loan guarantees, estimated to be worth $30 billion to the nuclear industry, for the construction of new nuclear power plants?

\_\_\_\_\_ YES

\_\_\_\_\_ NO

Comment:

4.  Do you support protection of the nuclear power industry from the full cost of liabilities due to accidents (e.g., as would be provided for by extension of the Price-Anderson act) or as the result of a terrorist attack?

_____ YES

_____ NO

Comment:


5.  Do you support the establishment of a high-level nuclear waste storage facility at Yucca Mountain, Nevada?

_____ YES

_____ NO

Comment:

# Notes

References to pages in *Time*, *Newsweek*, and *U.S. News & World Report* refer to the International Edition. Page numbers may be different in the edition for readers in the United States.

## Chapter 1

1. "Bottled Water." *The Ecologist*, February 2003, p. 46.

2. "EPA Says U.S. Economy Depends on Clean Water," CNN.com citation to *Nature*, June 9, 2000.

3. "Liquidating Our Assets." *World-Watch*, September-October 1997, p. 39.

4. S. Postel, *Last Oasis* (New York: Norton, 1992), 103.

5. N. L. Tomer, "Water Conservation Makes Dollars and Sense," 2001, http://www.crrs.net/story15.htm.

6. G. J. Kirmeyer, R. Williams, and P. D. Smith, *An Assessment of Water Distribution Systems and Associated Research Needs*, American Water Works Association Research Foundation report, Denver, 1994.

7. "Water Infrastructure Now: Clean and Safe Water in the 21st Century" (Washington, D.C.: Water Infrastructure Network, 2000).

8. Z. A. Smith, *The Environmental Policy Paradox*, 3rd ed. (Upper Saddle River, NJ: Prentice Hall, 2000), 130.

9. R. N. Stavins, "Price Water as Scarce, Vital Resource It Is," *Arizona Daily Star*, August 17, 1999, p. 13.

10. M. Margolis, "Drinking Water Goes Digital," *Newsweek*, August 27, 2001, p. 7.

11. A. P. McGinn, "Phasing Out Persistent Organic Pollutants," *State of the World 2000*, ed. L. Starke (New York: Norton, 2000), 79–100.

12. "News Briefs" *Environmental Science & Technology*, September 1997, p. 411A.

13. R. Showstack, "Survey of Emerging Contaminants in U.S. Indicates Need for Further Research," *EOS*, American Geophysical Union, March 26, 2002, p. 146.

14. H. I. Zeliger, "Toxic Effects of Chemical Mixtures," *Archives of Environmental Health*, January 2003, pp. 23–29.

15. K. S. Betts, "First Ecosystems Analysis Reveals Data Gaps," *Environmental Science & Technology*, November 1, 2002, p. 404A.

16. "An Environmental Report Card," *New York Times*, June 26, 2003, p. A3.

17. National Water Quality Inventory: 1998 Report to Congress. EPA 841-R-00-001 (Washington, D.C., EPA), http://www.epa.gov/305b.

18. P. Costner and J. Thornton, *We All Live Downstream: The Mississippi River and the National Toxics Crisis* (Washington, D.C.: Greenpeace, 1989).

19. Z. A. Smith, *The Environmental Policy Paradox*. 3rd ed. (Upper Saddle River, NJ: Prentice Hall, 2000), 185.

20. T. M. Kennedy and T. W. Lyons, "Hypoxia in the Gulf of Mexico: Causes, Consequences, and Political Considerations," *GSA Today* (Geological Society of America), August 2000, pp. 16–17.

21. Toxicspot.com, 1999.

22. "Military Pollution and Base Conversion," *Arc Ecology*, 2002, http://www.arc.home.igc.org/military.htm.

23. J. A. Kutner, *National Defense Magazine*, March 2000, http://www.nationaldefensemagazine.org/article.cfm?Id=52.

24. E. Bast, "U.S. and Canadian Water Pollution Jumps 26 Percent," *World-Watch*, September-October 2002, p. 11.

25. "Few Penalties for Permit Violators," *Chemical & Engineering News*, June 16, 2003, p. 19.

26. D. Shafer, "To Eat or Not to Eat," *E Magazine*, September-October 1999, p. 7.

27. "Facts About Pollution from Livestock Farms," Natural Resources Defense Council fact sheet, 2001, http://www.nrdc.org/water/pollution/ffarms.asp.

28. A. H. Smith, P. A. Lopipero, M. N. Bates, and C. M. Steinmaus, "Arsenic Epidemiology and Drinking Water Standards," *Science*, June 21, 2002, pp. 2145–2146.

29. J. Smutniak, "Living Dangerously: The Price of Prudence," *The Economist*, January 24, 2004, pp. 8–10.

## Chapter 2

1. University Corporation for Atmospheric Research, October 19, 2000, http://www.ucar.edu/communications/newsreleases/2000/floods.html.

2. Federal Emergency Management Agency, National Dam Safety Program, 1999, http://www.fema.gov/mit/ndspweb.htm.

3. "Trading Floods for Loess and Sinkholes," *Geotimes,* October 1994, p. 9.

4. J. Ackerman, "New Eyes on the Oceans," *National Geographic,* October 2000, p. 107.

5. R. Alonso-Zaldivar, "Flirting with Disaster," *Jerusalem Post,* September 9, 1998, p.8.

6. R. Alonso-Zaldivar, "Flirting with Disaster," *Jerusalem Post,* September 9, 1998, p.8.

7. J. H. Cushman, Jr., "Citing Insurance Costs, Group Urges Agencies to Buy More Houses in Flood Plains," *New York Times,* July 17, 1998, p. A15.

8. J. R. Minkel, "Political Watershed," *Scientific American,* January 2002, p. 21.

## Chapter 3

1. Environmental Protection Agency, 2003, http://www.epa.gov/epaoswer/non-hw/muncpl/fact.htm.

2. INFORM, http://www.informinc.org/GGTDraft.pdf.

3. "Short Takes," *International Herald Tribune,* April 6, 2000, p. 3.

4. "Disposable vs. Cloth," *Consumer Reports,* August 1998, p. 55.

5. Pennsylvania Environmental Network, http://www.penweb.org/issues/waste/importation/.

6. L. R. Brown, "New York: Garbage Capital of the World," *Earth Policy Institute News,* April 17, 2002, http://www.earth-policy.org/Updates/Update10.htm.

7. E. Helmore, "The Big Apple Sells Its Rotten Core," *The London Observer,* March 31, 1996, p. 21.

8. S. Battersby, "Plus c'est le Même Chews," *Nature,* February 20, 1997, p. 679.

9. "Short Takes," *International Herald Tribune,* April 28, 2000, p. 3.

10. "Uncovering Hazardous Waste Landfill Threats," *Environmental Science & Technology,* April 1, 2002, p. 139A.

11. C. Hogue, "Counting Down to Zero," *Chemical & Engineering News,* September 9, 2002, p. 31.

## Chapter 4

1. M. Freudenheim, "A Boom in Surgery to Shrink the Stomach," *International Herald Tribune,* August 30–31, 2003, p. 9; R. Winslow, "Obese American Teens Seek Stomach Surgery," *Wall Street Journal,* October 9, 2003, p. 20.

2. A. Ananthaswamy, "Cities Eat Away at Earth's Best Land," *New Scientist,* December 21/28, 2002, p. 9.

3. J. R. "Sprawling Over Croplands," *Science News,* March 4, 2000, p. 155.

4. M. Yudelman and J. M. Kealy, "The Graying Farmers," *Population Today*, May-June 2000, p. 6.

5. California Air Resources Board, "Smog and California Crops," 1991, pp. 2–3.

6. *Agriculture Fact Book 2001–2002*, U.S. Department of Agriculture, 2003, p. 24, http://www.usda.gov/factbook/2002factbook.pdf.

7. *Agriculture Fact Book 2001–2002*, U.S. Department of Agriculture, 2003, p. 30, http://www.usda.gov/nass/pubs/trends/farmnumbers.htm.

8. "Organic Agriculture: Implementing Ecology-Based Practices," *Organic Trade Association Newsletter*, No. 19, October-November 2001, p. 3.

9. Reuters, July 4, 2002. http://www.planetark.org/dailynewsstory.cfm.newsid/16699/story.

10. T. C. Rembert, "Food Porn," *E Magazine*, May-June, 1998, p. 19.

11. E. Dooley, "Protected Harvest," *Environmental Health Perspectives*, May 2002, p. A237.

12. H. I. Zeliger, "Toxic Effects of Chemical Mixtures," *Archives of Environmental Health*, January 2003, pp. 23–29.

13. D. J. Epstein, "Secret Ingredients," *Scientific American*, August 2003, p. 12.

14. B. Halweil, "Pesticide-Resistant Species Flourish," *Vital Signs*, ed. L. Starke (New York: Norton, 1999), 124.

15. "In Brief," *Environment*, September 2001, p. 8.

16. C. Bright, "Bioinvasions," *World-Watch*, July-August, 1998, p. 39.

17. Japan Chemical Week, December 21–28, 1995.

18. C. Green and A. Kremen, "U.S. Organic Farming in 2000–2001," *USDA Economic Research Service, Agriculture Information Bulletin no. 780*, 2003.

19. C. Green and A. Kremen, "U.S. Organic Farming in 2000–2001," *USDA Economic Research Service, Agriculture Information Bulletin* No. 780, 2003.

20. S. Deneen, "Food Fight," *E Magazine*, July-August 2003, p. 28.

21. B. Halweil, "Organic Food Found to be Higher in Heath-Promoting Compounds," *World-Watch*, July-August 2003, p. 9.

22. B. Halweil, "Organic Gold Rush," *World-Watch*, May-June 2001, p. 24.

23. R. Sheer, "Organic Profits," *E Magazine*, July-August 2003, p. 44, 46.

24. B. Halweil, "Farming in the Public Interest," *State of the World 2002*, ed. L. Starke (New York: Norton, 2002) p. 58.

25. K. Brown, "Seeds of Concern," *Scientific American*, April 2001, pp. 39–45.

26. "GM Food and Safety," *The Ecologist*, April 2003, p. 11.

27. J. Geary, "Risky Business," *Time*, July 28, 2003, p. 42.

28. "World Updates," *World-Watch*, July-August 2003, p. 11.

29. B. H. "Biotech Crop Laws Were Big in 2001," *Science News*, February 2, 2002, p. 77.

30. "Increased Plant Density in Northern Latitudes," *Environmental Science & Technology*, November 1, 2001, p. 443A.

## Chapter 5

1. Energy Information Administration World Energy Outlook 2003, http://www.eia.gov/oiaf/ieo/figure_12.html.

2. V. Smil, "Energy in the Twentieth Century: Resources, Conversions, Costs, Uses, and Consequences," *Annual Review of Energy and the Environment* 25 (2000):37.

3. A. Frangis, "Level Utility Billing Keeps a Lid on Costs," *Wall Street Journal*, June 8, 2003, p. 9.

4. M. Talwani, "Will Calgary Be the Next Kuwait?" *New York Times*, August 14, 2003; A. Tullo, "A New Source," *Chemical & Engineering News*, August 25, 2003, pp. 16–17.

5. CNBC/*Wall Street Journal*, http://www.msnbc.com/news/423165.asp?cpl=1.

6. C. J. Levine, "Jerusalem Blackout?" *Jerusalem Post*, September 5, 2003, p. 9.

7. B. Holmes and N. Jones. "Brace Yourself for the End of Cheap Oil," *New Scientist*, August 2, 2003, p. 8–11.

8. Energy Information Administration, http://tonto.eia.doe.gov/dnav/ng/hist/n3010us3M.htm.

9. R. Gold and B. Bahnea, "As Oil Costs Increase, LNG Market Expands," *Wall Street Journal*, July 18, 2003, p. A14.

10. P. Warburg, "That Fresh Air Can Kill You," *Jerusalem Post*, February 12, 2003, p. 7.

11. J. Johnson, "Blowing Green," *Chemical & Engineering News*, February 24, p. 28.

12. "Global Oil Production Falls as Gas Grows; China Burns Lots More Coal," *Energy World*, September 2003, p. 6; S. B. "Booming Economy, Booming Emissions," *Environment*, December 2003, p. 4.

13. J. Dunn, "Decarbonizing the Energy Economy," in *State of the World 2001*, ed. L. Starke (New York: Norton, 2001), 94–95.

14. J. Sawin, "Charting a New Energy Future," in *State of the World 2003*, ed. L. Starke (New York: Norton, 2003), 85–109.

15. Energy Information Administration, *Renewable Energy Annual 2002*, http://eia.doe.gov/cneaf/solar.renewables/page/rea_data/rea_sum.html.

16. Uranium Information Centre Nuclear Issues Briefing Paper No. 75, p. 2, http://www.uic.com.au/nip75.htm.

17. "Nuclear Safety Cracking," *The Ecologist*, December 2002-January 2003, p. 9.

18. J. L. Sawin, "European Wind Energy Production Reaches New Highs," *World-Watch*, May-June 2003, p. 8.

19. J. Johnson, "New Life for Nuclear Power?" *Chemical & Engineering News,* October 2, 2000, pp. 39–42.

20. D. M. Chapin, K. P. Cohen, W. K. Davis, E. E. Kinter, L. J. Koch, J. W. Landis, and M. Levinson et al., "Nuclear Power Plants and Their Fuel as Terrorist Targets," *Science,* September 20, 2002, pp. 1997–1999; P. Bunyard, "The Plane Truth," *The Ecologist,* November 2001, pp. 48–50.

21. M. Brower, *Cool Energy* (Cambridge, Mass.: MIT Press, 1998), 134.

22. M. O. Sheehan, "Solar Cell Use Rises Quickly," in *Vital Signs 2002* (New York: Norton, 2002), 44.

23. P. Fairley, "Solar on the Cheap," *Technology Review,* February 2002, p. 51.

24. P. Fairley, "Solar on the Cheap," *Technology Review,* February 2002, p. 52.

25. Solaraccess.com News, http://www.solaraccess.com/news/story?storyid=4803&p=1.

26. J. Johnson, "Blowing Green," *Chemical & Engineering News,* Feb. 24, 2003, pp. 27–30.

27. Energy Information Administration, *Renewable Energy Annual 2002,* http://eia.doe.gov/cneaf/solar.renewables/page/rea_data/rea_sum.html.

28. V. Herzog, T. E. Lipman, J. L. Edwards, and D. M. Kammen, "Renewable Energy," *Environment,* December 2001, pp. 8–20.

29. D. H. Freedman, "Fuel Cells vs. the Grid," *Technology Review,* February 2002, pp. 40–47.

30. N. Rader, *Power Surge* (Washington D.C.: Public Citizen, 1989).

31. M. Lee, "Tread Lightly," *The Ecologist Report,* November 2001, p. 35.

## Chapter 6

1. C. Parmesan and G. Yohe, "A Globally Coherent Fingerprint of Climate Change Impacts across Natural Systems," *Nature,* January 2, 2003, p. 37.

2. M. Z. Cutajar, "The Climate Protectors Need U.S. Leadership and Ingenuity," *International Herald Tribune,* March 20, 2001, p. 8.

3. S. Dunn, "Carbon Emissions Reach New High," in *Vital Signs 2002* (New York: Norton, 2002), 52.

4. Clean Air Council, "Health Impacts of Electricity Generation," http://www.cleanair.org/Energy/energy/Impacts.html.

5. M. O. Sheehan, "Making Better Transportation Choices," in *State of the World,* ed. L. Starke (New York: Norton, 2001), 105.

6. O. Gillham, *The Limitless City* (Washington, D.C.: Island Press), 93, 127.

7. Z. A. Smith, *The Environmental Policy Paradox,* 3rd ed. (Upper Saddle River, NJ: Prentice Hall, 2002), 163.

8. T. Bode, "Coming soon, a Broad Challenge to Polluting Automakers," *International Herald Tribune*, January 23, 2001, p. 8.

9. S. B. "Booming Economy, Booming Emissions," *Environment*, December 2003, p. 4.

10. E. Schmitt, "American Dream Update: More and More Mortgages, Too," *International Herald Tribune*, August 7, 2001, p. 3.

11. W. L. Chameides and M. Bergin, "Soot Takes Center Stage," *Science*, September 27, 2002, p. 2214.

12. M. Z. Jacobson, "Strong Radiative Heating Due to the Mixing State of Black Carbon in Atmospheric Aerosols," *Nature*, February 8, 2001, p. 695.

13. A. Ananthaswamy, "Crunch Time for the SUV," *New Scientist*, March 8, 2003, pp. 12–13.

14. Sierra Club SUV Report, http://www.sierraclub.org/globalwarming/2:suvreport/suvthreat.asp.

15. K. Keating, "Product Safety—The SUV Debate," Tuck School of Business, Dartmouth College, May 8, 2003, http://www.dartmouth.edu/tuck/news/media/pr20030508_ethics.html.

16. Sheehan, "Making Better Transportation Choices," p. 117.

17. "From Greenhouse Gas to Precious Gem in One Easy Step," *New Scientist*, July 26, 2003, p.17.

18. "Coal Fires are Global Catastrophe," BBC News, http://www.news.bbc.co.uk/1/hi/in_depth/sci/_tech/2003/denver_2003/2759983.stm.

19. K. M. Reese, "Australia Flights Methane," *Chemical & Engineering News*, June 18, 2001, p. 104.

20. H. W. Bange, "It's Not a Gas," *Nature*, November 16, 2000, p. 301.

21. Environmental Media Services, 2003, http://www.ems.org/sprawl/sub2_sprawl.html.

22. Sierra Club, http://www.sierraclub.org/sprawl/factsheet.asp.

23. E. Stokstad, "River Flow Could Derail Crucial Ocean Current," *Science*, December 13, 2002, p. 2110.

24. F. Pearce, "And Now the Weather: Rain, Rain, Here to Stay," *New Scientist*, November 29, 1997, p. 28.

25. L. Miller and B. C. Douglas, "Mass and Volume Contributions to Twentieth-Century Global Sea Level Rise," *Nature*, March 25, 2004, p. 406.

26. "The Tide is Higher: Study Shows Rise of Atlantic Sea Level," *Bulletin of the American Meteorological Society*, February 2002, p. 170.

27. M. Spalding, "Danger on the High Seas," *Geographical Magazine*, February 2002, p. 15.

28. L. Barnes, "Once Again, Israel Gets Away with Spending U.S. Aid on West Bank Settlements," *Washington Report on Middle East Affairs*, October-November 1995, p. 32, http://www.wrmea.com/backissues/1095/9510032.htm.

29. S. P. Leatherman, B. C. Douglas, and J. L. LaBreque, "Sea Level and Costal Erosion Require Large-Scale Monitoring," *EOS* (American Geophysical Union) January 14, 2003, pp. 13, 16.

30. "Erosion Threatens U.S. Coastal Homes," *International Herald Tribune,* June 28, 2000, p. 3.

31. M. Fishetti, "Drowning New Orleans," *Scientific American,* October 2001, p. 78.

32. D. Cromwell, "Ocean Circulation," *New Scientist* (insert), 2000.

33. R. A. Kerr, "Sea Change in the Atlantic," *Science,* January 2, 2004, p. 35.

34. C. W. Petit, "Perilous Waters," *U.S. News & World Report,* April 1, 2002, pp. 64–65.

35. R. Kunzig, "Against the Current," *U.S. News & World Report,* June 2, 2003, pp. 34–35.

36. S. Begley, "The Mercury's Rising," *Newsweek,* December 4, 2000, p. 78.

37. "Warmer World, More Males," http://www.CNN.com/NATURE/9908/06/diary.planet.

38. E. Asimov, "For Wines, the Paradox of Global Warming," *New York Times,* August 6, 2003, p. F1–F2; "It's Hot as Hell but the Bubbly's Lovely," *New Scientist,* November 15, 2003, p.17.

39. "Most EU Countries Will Fail on Renewables Targets," *Energy World,* January 2003, p. 7; "EU Behind Kyoto Schedule," *Environmental Science & Technology,* July 1, 2003, p. 247A.

40. Brookings Institute, "Greenhouse Gas Emissions," Policy Brief No. 52, http://www.Brookings.edu/dybdocroot/comm/PolicyBriefs/Pb052/pb52.htm.

41. "Estimated Emissions of Greenhouse Gases 1980–2001," Energy Information Administration, U.S. Department of Energy, 2001, http://www.eia.doe.gov/emeu/aer/txt/ptb1201.html.

42. M. O. Sheehan, "Carbon Emissions," in *Vital Signs 2003,* ed. L. Starke (New York: Norton, 2003), 41.

43. "Is Emission Trading Effective?" *Chemistry in Britain,* May 2003, p. 17.

## Chapter 7

1. Earth Policy Institute, September 17, 2002, http://www.earth-policy.org/Updates/Update17.htm.

2. American Lung Association, *Trends in Air Quality,* 2002, p. 2.

3. "Death in the City," *New Scientist,* April 13, 1996, p. 19.

4. "Traffic Fumes Lower Sperm Counts," *The Ecologist,* July-August 2003, p. 9.

5. P. Brinblecombe, "Acid Drops," *New Scientist* (insert), May 18, 2002, p. 2.

6. M. Higgins, "China Leads World in Sulfur Emissions," Environmental News Network, 1999, http:/www.enn.com/enn-news-archive/1999/11/112699/chinasulphur_7655.asp.

7. American Lung Association, *Trends in Air Quality*, 2002, p. 3.

8. J. Pelley, "Adirondack Lakes Recovering from Acid Rain," *Environmental Science & Technology*, June 1, 2003, pp. 202–203A.

9. American Lung Association, *Trends in Air Quality*, 2002, p. 15.

10. American Lung Association, *Trends in Air Quality*, 2002, pp. 15, 20.

11. F. Pearce, "False Forecasts Leave Cities Choking," *New Scientist*, February 8, 1997, p. 5.

12. Asthma and Allergy Foundation of America, http:/www.aafa.org/templ/ display .cfm?id=2&sub=25.

13. D. C., "Is a Faster Commute Worth It?" *Science News*, November 9, 2002, p. 302.

14. J. Selim, "Fetuses Take Air Pollution to Heart," *Discover*, April 2002, p. 12.

15. American Lung Association, *Trends in Air Quality*, 2002, p.6

16. F. Pearce, "Big City Killer," *New Scientist*, March 9, 2002, p. 8.

17. E. Nagourny, "Ricks from Sunscreen to Strokes," *International Herald Tribune*, December 18, 2003, p. 10.

18. American Lung Association, *Trends in Air Quality*, 2002, p. 8

19. F. Pearce, "Burning Rubber," *New Scientist*, April 10, 1999, p. 14.

20. R. M., "Florida Air Loaded with African Dust," *Science News*, June 14, 1997, p. 373.

21. P. H. Abelson, "Airborne Particulate Matter," *Science*, September 11, 1998, p. 1609.

22. O. B. Toon, "How Pollution Suppresses Rain," *Science*, March 10, 2000, pp. 1763–1764.

23. "News Briefs," *Environmental Science and Technology*, December 1996, p. 531A.

24. "Planes and Pollution," *Environmental Health Perspectives*, February 1997, p. 172.

25. R. Long, "Where There's Smoke, There's Pollution," *New York Times*, February 21, 2004, p. A15.

26. Environmental Protection Agency, *Latest Findings on National Air Quality: 2000 Status and Trends*, 2001, p. 1, http://www.epa.gov/oar/aqtrnd00/index .html.

27. P. Douglass, "The Chemicals in Cigarette Smoke," *Jerusalem Post*, August 17, 2003, http://pauldouglass.co.uk/stop_smoking_hypnotherapy/ chemicalsincigarettesmoke.htm.

28. E. Susman, "Smoking Inflames Joints," *Environmental Health Perspectives,* December 2002, p. A741.

29. K. H. Kilburn, "Stop Inhaling Smoke: Prevent Coronary Heat Disease," *Archives of Environmental Health,* February 2003, p. 71; American Lung Association, *Trends in Air Quality,* 2002, p. 9.

30. J. Siegel, "WHO Denounces Study on Passive Smoking," *Jerusalem Post,* March 11, 1998, p. 3.

31. "Costly Habit," *New Scientist,* April 5, 1997, p. 13.

32. N. Seppo, "Secondary Smoke Carries High Price," *Science News,* January 17, 1998, p. 36.

33. "Secondhand Smoke and Heart Disease," *Environmental Health Perspectives,* August 1997, pp. 788–789.

34. "A New Reason Not to Smoke," *Environmental Health Perspectives,* October 1997, p. 1044.

35. J. Siegel-Itzkovich, "Health Scan," *Jerusalem Post,* August 2, 1998.

36. "Airline is Ruled Liable in Asthma Death," *International Herald Tribune,* February 25, 2004, p. 4.

37. A. V. Nero, Jr., "Controlling Indoor Air Pollution," *Scientific American,* May 1988, p. 46.

38. "Radon—An Environmental Risk Caused Mainly by Nature," http://limankoski./porvoo.fi/comenius_english/products/radon1.htm.

39. Nero, "Controlling Indoor Air Pollution," p. 45.

40. K. S. Brown, "Sick Days at Work," *Environmental Health Perspectives,* October 1996, pp. 1032–1035.

41. Brown, "Sick Days at Work," p. 1032.

42. "Chemical Warfare," *WorldWatch,* July-August 2003, p. 32.

## Chapter 8

1. "A Spectrum of Skin Protection," *Consumer Reports,* May 1998, pp. 20–23.

2. J. Dini, "CFC Treaty Fuels Black Market," Heartland Institute, 2000, http://www.heartland.org/Article.cfm?artID=9614.

3. F. Pearce. "A Very Bad Atmosphere," *New Scientist,* September 20, 1997, p. 12.

4. D. T. Shindett, D. Rind, and P. Lonergan, "Increased Stratospheric Losses and Delayed Eventual Recovery Owing to Increased Greenhouse Gas Concentrations," *Nature,* April 9, 1998, p. 589.

5. E. H. Steinberger, personal communication, 1999.

6. D. Friedman, "Sun Safety," *Jerusalem Post* (insert), May 7, 1999, p. 11.

7. National Weather Service, "The Ultraviolet Index," 1995, http://www.nws.noaa.gov/pa/secnews/uv/uv1.htm.

8. "Away from Politics," *International Herald Tribune,* May 1999, pp. 15–16.

9. "How Many People Get Melanoma Skin Cancer?" American Cancer Society, http://www./CRI_2_2_1X_How_many_people_get_melanoma_skin_cancer_50.asp?sitearea=CRI&viewmode=prin.

10. N. Seppa, "Skin Cancer Makes Unexpected Reappearance," *Science News,* June 21, 1997, p. 383.

11. National Weather Service, "The Ultraviolet Index," 1995, http://www.nws.noaa.gov/ pa/secnews/uv/uv1.htm.

12. M. Melton, "Deep-Fried by the Sun," *U.S. News & World Report,* September 14, 1998, p. 50.

13. Texas Cooperative Extension, 2002, http://fes.tamu.edu/health/cancer/indoortan/tanning.htm.

14. "Unhealthy Glow," *U.S. News & World Report,* October 27, 2003, p. 16.

15. N. Izenberg, "Kids Health for Parents," http://www.kidshealth.org/parent/question/safety/tanning_booths.html.

16. "HealthLink," Medical College of Wisconsin, 2002, http://healthlink.mcw.edu/article/964647970.html.

17. C. Biever, "Bring Me Sunshine," *New Scientist,* August 9, 2003, p. 30.

## Chapter 9

1. "State of the Planet," *The Ecologist,* September 2002, p. 9.

2. "Nuclear Power Plants Shut," *The Ecologist,* July-August 2002, p. 8.

3. *Counter Punch,* January 1–15, 1999, http://www.21stcenturyradio.com/NP7-24–99–5.html; http://archive.greenpeace.org/majordomo/index-press-releases/1998/msg00077.html; http://www.radtexas.org/corpresponsibility.html; http://www.ananova.com/news/story/sm_768980.html?menu=.

4. H. Edwards, "Radioactive Blunders Double in a Decade," *New Scientist,* February 9, 2002, p. 6.

5. Z. A. Smith, *The Environmental Policy Paradox,* 3rd ed. (Upper Saddle River, NJ: Prentice Hall, 2002), 154.

6. "Climate Change's Genetic Effect," *The Ecologist,* April 2003, p. 9.

7. P. Bunyard, "The Plane Truth," *The Ecologist,* November 2001, p. 48.

8. "UK: The Tooth Fairy Comes to Britain," *The Ecologist* May 2000, p. 14.

9. P. Bunyard, "Making a Mess of MOX," *The Ecologist,* February 2002, p. 67.

10. K. A. Svital, "Children of Chernobyl," *Discover,* August 2001, p. 10.

11. W. Drozdiak, "40 Nations Fund New Chernobyl Safety Shell," *International Herald Tribune,* July 6, 2000, pp. 1, 3.

12. J. Litvin, "Chernobyl's Cry Grows Louder," *Jerusalem Post,* September 11, 2000, p. 6.

13. Ibid.

14. "Sound Bites," *The Ecologist*, December 2002-January 2003, p. 8.

15. R. Edwards, "Amorous Worm Reveal the Effect of Chernobyl Fallout on Wildlife," *New Scientist*, April 12, 2003, p. 10.

16. E. Lukatsky, "Inside the Sarcophagus," *Jerusalem Post*, December 12, 2000, p. 7.

17. S. LaFraniere, "Last Chernobyl Reactor Shut Down," *International Herald Tribune*, December 16–17, 2000, pp. 1, 4.

18. M. Wines, "Russia and Ukraine Split on Chernobyl Risk," *International Herald Tribune*, April 24, 2003, p. 4.

19. "Chernobyl Officials Admit Danger of Sarcophagus Collapse," *Jerusalem Post*, April 25, 2003, p. A7.

20. "Special Report: Could It Happen to Us? Electricity in Other Countries," *The Economist*, August 23–29, 2003, p. 19.

21. "Nuclear Safety," *New Scientist*, June 14, 2003, p. 9.

22. G. Wehrfritz and A. Webb, "Breach of Faith," *Newsweek*, September 30, 2002, pp. 41–43.

23. R. Edwards, "Are Aging U.S. Reactors Safe?" *New Scientist*, August 9, 2003, p. 13.

24. "Nuclear Waste," *National Geographic*, September 2002, p. 109.

25. J. Wheelwright, "Welcome to Yucca Mountain," *Discover*, September 2002, p. 70.

26. Department of Energy, http://enr.construction.com/news/powerindus/archives/010507.asp.

27. J. Johnson, "Green Light for Yucca Mountain," *Chemical & Engineering News*, January 21, 2002, p. 9.

28. J. Johnson, "Bush Approves Yucca Mountain," *Chemical & Engineering News*, February 25, 2002, p. 8.

29. C. L. Grossman, R. H. Nussbaum, and F. D. Nussbaum, "Thyrotoxicosis among Hanford, Washington, Downwinders: A Community-Based Health Survey," *Archives of Environmental Health*, January-February 2002, pp. 9–15; C. M. Grossman, R. H. Nussbaum, and F. D. Nussbaum, "Cancers among Residents Downwind of Hanford, Washington, Plutonium Production Site," *Archives of Environmental Health*, May 2003, pp. 267–274.

30. K. D. Crowley and J. F. Ahearne, "Managing the Environmental Legacy of U.S. Nuclear Weapons Production," *American Scientist*, November-December 2002, pp. 514–523; J. Johnson, "Hanford on Fast-Forward," *Chemical & Engineering News*, June 10, 2002, pp. 24–33; S. Kershaw and M. L. Wald, "Lack of Safety is Charged in Nuclear Site Cleanup," *New York Times*, Feb. 20, 2004, pp. A1, A22.

# Chapter 10

1. Environmental Working Group, "Fuel Efficient Cars Could Save American Households $590 a Car at the Pump and Slash Global Warming Pollution," 1997; Testimony by L. G. Billings to Senate Subcommittee, 1997; T. Paine, Common Sense, 2000; Brainy Quote, 2004.

2. B. E. Erickson, "The President's 2004 Budget," *Environmental Science & Technology*, April 1, 2003, pp. 128A–129A.

3. R. Kennedy, "The Day The Traffic Disappeared," *New York Times Magazine*, April 20, 2003, pp. 42–45.

4. M. Hamer, "I'm a Commuter, Get Me Out of Here," *New Scientist*, November 8, 2003, p. 36.

5. Chevron Corporation, http://www.chevron.com/about/currentissues/gasoline/apiprice/gasoline_price_trends.shtm.

6. "Massachusetts Environmental Taxes," http://www.environmentalleague.org/Issues/Tax_Shifting/environmental_taxes.htm; N. Goodman and A. Reynolds, "Environmental Tax Shifting in Massachusetts," 2001, http://www.environmentalleague.org/primer.pdf.

7. European Environment Agency, 2000, http://reports.eea.eu.int/Environmental_Issues_No_18/en/envissue18.pdf.

# Additional Readings

References to pages in *Time, Newsweek,* and *U.S. News & World Report* refer to the International Edition. Page numbers may be different in the edition for readers in the United States.

For a guide to worldwide environmental organizations, see the 2003 *Conservation Directory* compiled by the National Wildlife Federation and published by Island Press, Washington, D.C.

## Introduction

Greenberg, M. R. Is Public Support for Environmental Protection Decreasing? An Analysis of U.S. and New Jersey Data. *Environmental Health Perspectives,* February 2004, pp. 121–125.

Guber, D. L., *The Grassroots of a Green Revolution* (Cambridge, MA.: MIT Press), 2003.

## Chapter 1: Water: Is There Enough and Is It Drinkable?

Gleick, P. H., and S. Postel. Safeguarding Our Water. *Scientific American,* February 2001, pp. 26–41.

Glennon, R. *Water Follies: Groundwater Pumping and the Fate of America's Fresh Waters.* Washington, DC: Island Press, 2002.

Howard, B. Message in a Bottle. *E Magazine,* September-October 2003, pp. 26–39.

It's Only Water, Right? *Consumer Reports,* August 2000, pp. 17–21.

Lavelle, M., and J. Kurlantzick. The Coming Water Crisis. *U.S. News & World Report,* August 12, 2002, pp. 23–30.

Markels, A. Water Fights. *U.S. News & World Report,* May 19, 2003, pp. 58–61.

Markowitz, G., and D. Rosner. *Deceit and Denial: The Deadly Politics of Industrial Pollution.* Berkeley: University of California Press, 2002.

Postel, S. Redesigning Irrigated Agriculture. In L. Starke, ed., *State of the World 2000*, 39–58. New York: Norton, 2000.

Powers, S. Chemicals in water supplies produce feminizing effects, lower sperm quality. *Worldwatch*, Nov/Dec 2003, p. 11.

Russell, D. America's 13 Most Endangered Rivers: Can They Be Saved? *E Magazine*, May-June 2002, pp. 26–33.

Sampat, P. Uncovering Groundwater Pollution. In L. Starke, ed., *State of the World 2001*, 21–42. New York: Norton, 2001.

Shermer, M. Bottled Twaddle. *Scientific American*, July 2003, p. 23.

Sunstein, C. R. *Risk and Reason: Safety, Law, and the Environment*. New York: Cambridge University Press, 2002.

*The Economist*, Living Dangerously: The Price of Prudence. January 24, 2004, pp. 8–10.

U.S. Environmental Protection Agency. How to Conserve Water and Use It Effectively. http://www.epa.gov/water/you/chap3.html.

U.S. Geological Survey. *The Quality of Our Nation's Waters—Nutrients and Pesticides*. U.S. Geological Survey Circular 1225. Washington, DC: U.S. Geological Survey, 1999.

Whitman, D. The Sickening Sewer Crisis. *U.S. News & World Report*, June 12, 2000, pp. 16–18.

## Chapter 2: Floods: Too Much Water

Baish, S. K., S. D. David, and W. L. Graf. The Complex Decisionmaking Process for Removing Dams. *Environment*, May 2002, pp. 20–31.

Changnon, S. A., and D. R. Easterling. U.S. Policies Pertaining to Weather and Climate Extremes. *Science*, September 22, 2000, pp. 2053–2055.

Chrzastowski, M. J., M. M. Killey, R. A. Bauer, P. B. DuMontelle, A. L. Erdmann, B. L. Herzog, J. M. Masters, and L. R. Smith. *The Great Flood of 1993*. Illinois State Geological Survey Special Report 2. Champaign: Illinois State Geological Survey, 1994.

Downton, M. W., and R. A. Pielke, Jr. Discretion without Accountability: Politics, Flood Damage, and Climate. *Natural Hazards Review* 2 (2001): 157–166.

Doyle, M. W., E. H. Stanley, J. H. Harbor, and G. S. Grant. Dam Removal in the United States: Emerging Needs for Science and Policy. *EOS*, January 28, 2003, pp. 31–33.

H. John Heinz, III. Center for Science, Economics, and the Environment. *The Hidden Costs of Coastal Hazards*. Washington, DC: Island Press, 2000.

Jacobs, J. W., and J. L. Wescoat, Jr. Managing River Resources. *Environment*, March 2002, pp. 8–19.

Longman, P. J. By the Sea, Subsidies. *U.S. News & World Report*, October 4, 1999, pp. 38–40.

Lowry, W. R. Dam Politics: Restoring America's Rivers. Washington, DC: Georgetown University Press, 2003.

McCully, P. *Silenced Rivers—The Ecology and Politics of Large Dams*. London: Zed Books, 2001.

Middleton, N. Coming to Blows. *Geographical Magazine*, October 2000, pp. 54–58.

Pearce, F. We Can't Hold Back the Water any More. *New Scientist*, January 10, 2004, pp. 26–29.

Perry, C. A. Significant Floods in the United States during the 20th Century— USGS Measures a Century of Floods. USGS Fact Sheet 024–00, 2000. http://ks .water.usgs.gov/Kansas/pubs/fact-sheets/fs.024-00.html.

Riggs, S. R. Conflict on the Not-So-Fragile Barrier Islands. *Geotimes*, December 1996, pp. 14–18.

Sheaffer, J. R., J. D. Mullan, and N. B. Hinch. Encouraging Wise Use of Floodplains with Market-Based Incentives. *Environment*, January-February 2002, pp. 32–43.

Watson, B. A Town Makes History by Rising to New Heights. *Smithsonian*, June 1996, pp. 110–120.

## Chapter 3: Garbage: The Smelly Mountain

Environmental Protection Agency. *Municipal Solid Waste in the United States: 2000 Facts and Figures*. Washington, DC: Environmental Protection Agency, 2002.

Gardner, G. Recycling Organic Waste: From Urban Pollutant to Farm Resource. Worldwatch Paper 135. Washington, DC: Worldwatch Institute, 1997.

Grove, N. Recycling. *National Geographic*, July 1994, pp. 92–115.

A Guilt-Free Guide to Garbage. *Consumer Reports*, February 1994, pp. 91–114.

Hasan, S. E. *Geology and Hazardous Waste Management*. Upper Saddle River, NJ: Prentice Hall, 1996.

Motavalli, J. Zero Waste. *E Magazine*, March-April 2001, pp. 26–33.

National Science Foundation. *Understanding Garbage and Our Environment*. New York: McGraw-Hill, 1998.

Pellerin, C. Alternatives to Incineration: There's More Than One Way to Remediate. *Environmental Health Perspectives* 102, no. 10 (1994): 840–845.

Probst, K. N., and D. M. Konisky. *Superfund's Future: What Will It Cost?* Washington, DC: RFF Press, 2001.

Rathje, W. L. Once and Future Landfills. *National Geographic*, May 1991, pp. 116–134.

Steinberg, T. Walking Sewers. *Natural History,* December 2002/January 2003, p. 80.

Taylor, D. Talking Trash. *Environmental Health Perspectives* 107 (1999): A404–A409.

Trinh, J. Energy from Trash Tires. *E Magazine,* July-August 2002, p. 10.

Whitman, D. The Sickening Sewer Crisis. *U.S. News & World Report,* June 12, 2000, pp. 16–18.

## Chapter 4: Soil, Crops, and Food

Ackerman, J. How Altered? *National Geographic,* May 2002, pp. 33–50.

Beyond Organics. *New Scientist,* May 18, 2002, pp. 32–41.

Brown, K., K. Hopkin, and M. Mellon. Genetically Modified Foods: Are They Safe? *Scientific American,* April 2001, pp. 39–51.

Cohen, P. Fighting over Pharming. *New Scientist,* March 1, 2003, pp. 22–23.

Deneen, S. Food Fight. *E Magazine,* July-August 2003, pp. 26–33.

Eckley, N. Traveling Toxics. *Environment,* September 2001, pp. 24–36.

Glasser, J. A Broken Heartland. *U.S. News & World Report,* May 7, 2001, pp. 16–20.

Guterl, F. Brave New Foods. *Newsweek,* January 28, 2002, pp. 47–51.

Halweil, B. Farming in the Public Interest. *State of the World 2002,* 51–74. New York: Norton, 2002.

Halweil, B. Organic Gold Rush. *World-Watch,* May-June 2001, pp. 22–32.

Hileman, B. Drugs from Plants Stir Debate. *Chemical & Engineering News,* August 12, 2002, pp. 22–25.

How Safe Is Our Produce? *Consumer Reports,* March 1999, pp. 28–31.

Marcus, M. B. Organic Foods Offer Peace of Mind—At a Price. *U.S. News & World Report,* January 15, 2001, pp. 48–50.

Reilly, J., F. Tubiello, B. McCarl, D. Abler, R. Darwin, K. Fuglie, S. Hollinger, et al. U.S. Agriculture and Climate Change: New Results. *Climatic Change* 57 (2003): 43–69.

Warwick, H., and G. Meziani. Seeds of Doubt: North American Farmers' Experiences of GM Crops. Bristol, UK: Soil Association, 2002.

## Chapter 5: Energy Supplies

Benditt, J., ed. *Technology Review* (special issue on energy), February 2002.

Boyle, S. Renewable Energy. *Geographical Magazine,* August 2002, pp. 71–93.

British Petroleum. *Statistical Review of World Energy 2001.* London: British Petroleum, 2002. (Annual publication containing data about reserves, production, and consumption of oil, natural gas, hydropower, and nuclear energy for countries of the world.)

Craig, P. P., A. Gadgil, and J. G. Koomey. What Can History Teach Us? A Retrospective Examination of Long-Term Energy Forecasts for the United States. *Annual Review of Energy & Environment* 27 (2002): 83–118.

Dunn, S. Decarbonizing the Energy Economy. *State of the World 2001*, 83–102. New York: Norton, 2001.

Hartsook, C. *United States Energy Industry Overview.* Agricultural Marketing Resource Center, Iowa State University, 2003.

Herzog, A. V., T. E. Lipman, J. L. Edwards, and D. M. Kammen. Renewable Energy, A Viable Choice. *Environment*, December 2001, pp. 8–20.

Johnson, J. Blowing Green. *Chemical & Engineering News*, February 24, 2003, pp. 27–30.

Lavelle, M. Living without Oil. *U.S. News & World Report*, February 17, 2003, pp. 32–39.

Lavelle, M. Sand Dollars. *U.S. News & World Report*, October 13, 2003, pp. 34–36.

Pinsker, L. M. A Second Look at Geothermal Energy. *Geotimes*, June 2003, p. 31.

Rifkin, J. The Hydrogen Economy. *E Magazine*, January-February 2003, pp. 26–39.

Rifkin, J. The Hydrogen Economy: *The Creation of the Worldwide Energy Web and the Redistribution of Power on Earth.* New York: Jeremy P. Tarcher/Putnam, 2002.

Sang, D. Atoms Unleashed. *New Scientist* (insert), January 18, 2003.

Sawin, J. Charting a New Energy Future. *State of the World 2003*, 85–109. New York: Norton, 2003.

Smil, V. Energy in the Twentieth Century: Resources, Conversions, Costs, Uses, and Consequences. *Annual Review of Energy & Environment* 25 (2000): 21–51.

There's Oil in Them Thar Sands! *Economist*, June 28–July 4, 2003, pp. 91–92.

## Chapter 6: Global Warming: The Climate Is Changing

Dean, C. *Against the Tide: The Battle for America's Beaches.* New York: Columbia University Press, 1999.

Dunn, S., R. Friedman, and S. Baish. Coastal Erosion: Evaluating the Risk. *Environment*, September 2000, pp. 36–45.

Epstein, P. R. Is Global Warming Harmful to Health? *Scientific American*, August 2000, pp. 36–43.

Fischetti, M. Drowning New Orleans. *Scientific American*, October 2001, pp. 76–85.

Gelbspan, R. Reality Check. *E Magazine*, September-October 2000, pp. 24–39.

Guterl, R. The Truth about Global Warming. *Newsweek*, July 23, 2001, pp. 44–49.

Holdren, J. P. The Energy-Climate Challenge: Issues for the New U.S. Administration. *Environment,* June 2001, pp. 8–21.

Kates, R. W., and T. J. Wilbanks. Making the Global Local. *Environment,* April 2003, pp. 12–23.

Kluger, J., and M. D. Lemonick. Climate of Despair. *Time,* April 23, 2001, pp. 50–59.

Petit, C. W. Perilous Waters. *U.S. News & World Report,* April 1, 2002, pp. 64–65.

Rosenzweig, C., and W. D. Solecki. Climate Change and a Global City: Learning from New York. *Environment,* April 2001, pp. 8–18.

Rowland, S. Climate Change and its Consequences. *Environment,* March 2001, pp. 28–34.

Schulman, A. Insured Destruction. *E Magazine,* July-August, 2002, pp. 16–17.

Taylor, K. Rapid Climate Change. *American Scientist,* July-August 1999, pp. 320–327.

Trenberth, K. E. Stronger Evidence of Human Influences on Climate: The 2001 IPCC Assessment. *Environment,* May 2001, pp. 8–19.

Walther, G-R., E. Post, P. Convey, A. Menzel, C. Parmesan, T. J. C. Beebee, J-M. Fromefitin, O. Hoegh-Guldberg, and F. Bairlein. Ecological Responses to Recent Climate Change. *Nature,* March 28, 2002, pp. 389–395.

Wilbanks, T. J., S. M. Kane, P. N. Leiby, R. D. Perlack, C. Settle, J. F. Shogren, and J. B. Smith. Possible Responses to Global Climate Change: Integrating Mitigation and Adaptation. *Environment,* June 2003, pp. 28–38.

## Chapter 7: Air Pollution and Your Lungs

American Lung Association. *Trends in Air Quality.* New York: American Lung Association, 2002.

Barrett, J. R. Mold Insurance. *Environmental Health Perspectives.* February 2003, pp. A100–A103.

Brimblecombe, P. Acid Drops. *New Scientist* (insert), May 18, 2002.

Cifuentes, L., V. H. Borja-Aburto, N. Gouveia, G. Thurston, and D. L. Davis. Hidden Health Benefits of Greenhouse Gas Mitigation. *Science,* August 17, 2001, pp. 1257–1259.

Cole, L. A. *Element of Risk: The Politics of Radon.* Washington, DC: 1993. American Association for the Advancement of Science Press.

Davis, D. *When Smoke Ran Like Water: Tales of Environmental Deception and the Battle Against Pollution.* Basic Books, New York, 2002.

Ellis, R. Secondhand Smoke and the Helena Story. *International Herald Tribune,* October 17, 2003, p. 5.

Fenger, J. Urban Air Quality. *Atmospheric Environment* 33 (1999): 4877–4900.

Frazer, L. Seeing through Soot. *Environmental Health Perspectives*, August 2002, pp. A470–A473.

Harrington, W., and V. McConnell. A Lighter Tread. *Environment*, November, 2003, pp. 22–39.

Holzman, D. Plane Pollution. *Environmental Health Perspectives* 105 (1997): 1300–1305.

Jones, A. P. Indoor Air Quality and Health. *Atmospheric Environment* 33 (1999): 4535–4564.

Koren, H. S., and Utell, M. J. Asthma and the Environment. *Environmental Health Perspectives* 105 (1997): 534–537.

Manuel, J. S. Dedicated Outdoor Air Systems: Rx for Sick Buildings. *Environmental Health Perspectives*, October, 2003, pp. A712–A715.

Merefield, J. Dust to Dust. *New Scientist* (insert), September 21, 2002.

Ott, W. R., and J. W. Roberts. Everyday Exposure to Toxic Pollutants. *Scientific American*, February 1998, pp. 72–77.

Wakefield, J. The Lead Effect. *Environmental Health Perspectives*, October 2002, pp. A574–A580.

## Chapter 8: Skin Cancer and the Ozone Hole

Anderson, I. Sun Worshippers Pay with Their Skin. *New Scientist*, December 21, 1996, p. 9.

Baker, L. The Hole in the Sky. *E Magazine*, November-December 2000, pp. 34–39.

deBlanc-Knowles, J., and K. Allen. Dangerous Tans. *E Magazine*, September-October 2003, pp. 11–12.

De Gruijl, F. R. Impacts of a Projected Depletion of the Ozone Layer. *Consequences* 1, no. 2 (1995): 12–21.

Demers, S., S. Roy, and S. de Mora. The Impact of Ozone Layer Depletion on the Marine Environment. *Ecodecision*, winter 1996, pp. 67–70.

*Journal of Photochemistry and Photobiology B. Biology.* Environmental Effects of Ozone Depletion, 1998 Assessment. (Seven articles on the effects of ozone depletion on human health, terrestrial and aquatic ecosystems, and other things.)

Kane, R. P. Ozone Depletion, Related UVB Changes and Increased Skin Cancer Incidence. *International Journal of Climatology* 18 (1998): 457–472.

Kerlin, K. Blocking the Burn. *E Magazine*, July-August 2001, pp. 52–55.

Marks, R. *Sun and the Skin.* 2nd ed. London: Martin Dunitz, 1995.

Royal Swedish Academy of Sciences. Environmental Effects of Ozone Depletion. *Ambio* (special issue), 24, no. 3 (1995): 137–196.

A Spectrum of Sun Protection. *Consumer Reports*, May 1998, pp. 20–23.

Yokouchi, Y., D. Toom-Sauntry, K. Yazawa, T. Inagaki, and T. Tamaru. Recent Decline of Methyl Bromide in the Troposphere. *Atmospheric Environment* 36 (2002): 4985–4989.

## Chapter 9: Nuclear-Waste Disposal: Not In My Backyard

Crowley, K. D., and J. F. Ahearne. Managing the Environmental Legacy of U.S. Nuclear-Weapons Production. *American Scientist,* November-December 2002, pp. 514–523.

Farber, D., and J. Weeks. Graceful Exit? Decommissioning Nuclear Reactors. *Environment,* July-August 2001, pp. 8–21.

Gaines, M. Radiation and Risk. *New Scientist* (insert), March 18, 2000.

Hanks, T. C., I. J. Winograd, R. E. Anderson, T. E. Reilly, and E. P. Weeks. *Yucca Mountain as a Radioactive-Waste Repository.* U.S. Geological Survey Circular 1184. Washington, DC: U.S. Geological Survey, 1999.

Johnson, J. Hanford on Fast-Forward. *Chemical and Engineering News,* June 10, 2002, pp. 24–33.

Johnson, J. Roads or Rails. *Chemical & Engineering News,* January 4, 2003.

Johnson, J. Who's Watching the Reactors? *Chemical & Engineering News,* May 12, 2003, pp. 27–31.

Interdisciplinary Science Reviews. *Nuclear Power in the Twenty-First Century* 24, no. 4 (winter 2001).

Macfarlane, A. Interim Storage of Spent Fuel in the United States. *Annual Review of Energy & Environment* 26 (2001): 201–231.

O'Hanlon, L. The Time-Travelling Mountain. *New Scientist,* July 1, 2000, pp. 30–33.

Stone, R. Living in the Shadow of Chernobyl. *Science,* April 20, 2001, pp. 420–426.

Wald, M. L. Dismantling Nuclear Reactors. *Scientific American,* March 2003, pp. 36–45.

Wehrfritz, G., and A. Webb. Breach of Faith. *Newsweek,* September 30, 2002, pp. 40–42.

Zorpette, G. Hanford's Nuclear Wasteland. *Scientific American,* May 1996, pp. 72–81.

## Chapter 10: Conclusion

Alexander, C. P., ed. Our Precious Planet. *Time* (special issue), November 1997.

Ausubel, J. H., D. G. Victor, and I. K. Wernick. The Environment Since 1970. *Consequences* 1, no. 3 (1995): 2–15.

Begley, S. The Battle for Planet Earth. *Newsweek,* April 24, 2000, pp. 47–49.

Brower, M., and W. Leon. *The Consumer's Guide to Effective Environmental Choices.* New York: Three Rivers Press, 1999.

Brown, L. R., and J. Mitchell. Building a New Economy. *State of the World 1998,* 168–187. New York: Norton, 1998.

Chess, C., and K. Purcell. Public Participation in the Environment: Do We Know What Works? *Environmental Science & Technology* 33, no. 16 (1999): 2685–2691.

Dunkiel, B., M. J. Hamond, and J. Motavalli. Sharing the Wealth. *E Magazine,* March-April 1999, pp. 28–35.

Easterling, D. R., G. A. Meehl, C. Parmesan, S. A. Changnon, T. R. Karl, and L. O. Mearns. Climate Extremes: Observations, Modeling, and Impacts. *Science,* September 22, 2000, pp. 2068–2074.

Environmental League of Massachusetts. *Environmental Tax Shifting in Massachusetts.* Boston: Environmental League of Massachusetts, 2001.

European Environment Agency. *Environmental Taxes: Recent Developments in Tools for Integration.* Copenhagen: European Environment Agency, 2000.

Hoffert, M. I., K. Caldeinra, G. Benford, D. R. Criswell, C. Green, H. Herzog, et al. Advanced Technology Paths to Global Climate Stability: Energy for a Greenhouse Planet. *Science,* November 1, 2002, pp. 981–987.

Roodman, D. M. Continental Tax Shifts Multiplying. In L. Starke, ed. *Vital Signs 2000,* 138–141. New York: Norton, 2000.

Roodman, D. M. Getting the Signals Right: Tax Reform to Protect the Environment and the Economy. Worldwatch Paper 134. Washington, DC: Worldwatch Institute, 1997.

Turner, C., and E-A. Zen. Engaging "My Neighbor" in the Issue of Sustainability, Part IX: We Live in a World of Change. *GSA Today,* September 2000, p. 9.

## General Readings

Annan, K. A. Sustaining the Earth in the New Millennium. *Environment,* October 2000, pp. 20–30.

Bailey, R. *Earth Report 2000.* New York: McGraw-Hill, 2000.

Beder, S. *Global Spin: The Corporate Assault on Environmentalism.* White River Junction, VT: Chelsea Green Publishing, 1997.

Blatt, H. *Our Geologic Environment.* Upper Saddle River, NJ: Prentice Hall, 1997.

Coonie, C. M. Still Searching for Environmental Justice. *Environmental Science & Technology,* May 1, 1999, pp. 200A–204A.

Farrow, S., and M. Toman. Using Benefit-Cost Analysis to Improve Environmental Regulations. *Environment,* March 1999, pp. 12–15, 33–38.

Goudie, A., and H. Viles. *The Earth Transformed: An Introduction to Human Impacts on the Environment.* Oxford, UK: Blackwell, 1997.

Hardin, G. *The Ostrich Factor: Our Population Myopia.* New York: Oxford University Press, 1999.

McKibben, B. *The End of Nature.* New York: Random House, 1989.

McNeill, J. R. *Something New under the Sun: An Environmental History of the Twentieth Century.* New York: Norton, 2001.

Motavalli, J. Getting Out the Vote. *E Magazine,* May-June 2003, pp. 26–33.

Our Precious Planet. *Time* (special issue), November 1997.

Roper Starch Worldwide. *The National Report Card on Environmental Knowledge Attitudes and Behaviors.* Washington, DC: National Environmental Education and Training Foundation, 1999.

Smith, Z. A. *The Environmental Policy Paradox.* 3rd ed. Upper Saddle River, NJ: Prentice Hall, 2000.

Wagner, L. A. *Materials in the Economy—Material Flows, Scarcity, and the Environment.* U.S. Geological Survey Circular 1221. Washington, DC: U.S. Geological Survey, 2002.

Worldwatch Institute, annually since 1984. State of the World. A series of books, each containing about 10 articles on environmentally important topics, written by experts. New York: Norton, 1984–.

Worldwatch Institute, annually since 1992. Vital Signs. A series of books tracking worldwide environmental trends. New York: Norton, 1992–.

Zimmerman, M. *Science, Nonscience, and Nonsense: Approaching Environmental Literacy.* Baltimore: Johns Hopkins University Press, 1995.

    Magazines accessible to nonscientists that regularly contain articles about environmental problems include the following:

*The Ecologist.* 10 issues/year. www.theecologist.org.

*E: The Environmental Magazine,* 6 issues/year. www.emagazine.com.

*Environment.* 10 issues/year. www.heldref.org.

*New Scientist.* Weekly. www.newscientist.com

*World-Watch.* 6 issues/year. www.worldwatch.org.

# Index